REAL MARKETS AND ENVIRONMENTAL CHANGE IN KERALA, INDIA

To Lori

Real Markets and Environmental Change in Kerala, India

A new understanding of the impact of crop markets on sustainable development

RENÉ VÉRON
Research Fellow, Department of Geography, University of Zurich and Academic Visitor, School of Oriental and African Studies, University of London

LONDON AND NEW YORK

First published 1999 by Ashgate Publishing

Reissued 2018 by Routledge
2 Park Square, Milton Park, Abingdon, Oxon, OX14 4RN
711 Third Avenue, New York, NY 10017, USA

Routledge is an imprint of the Taylor & Francis Group, an informa business

Publisher's Note
The publisher has gone to great lengths to ensure the quality of this reprint but points out that some imperfections in the original copies may be apparent.

Disclaimer
The publisher has made every effort to trace copyright holders and welcomes correspondence from those they have been unable to contact.

A Library of Congress record exists under LC control number: 99073397

ISBN 13: 978-1-138-32907-2 (hbk)
ISBN 13: 978-1-138-32909-6 (pbk)
ISBN 13: 978-0-429-44832-4 (ebk)

Contents

List of Figures

List of Tables

Acknowledgements

Many people have helped me in the process of writing and editing this book, and preparing and conducting the corresponding study. This book draws on my earlier Ph.D. thesis "Markets, Environment and Development in South India: Cultivation and Marketing of Pineapple and Cashew in Kerala" submitted to the University of Zurich, Faculty of Science (Philosophische Fakultät II) in January 1998. I am indebted to Prof. Dr. Ulrike Müller-Böker, Head of the Human Geography Division, Department of Geography, University of Zurich, and to her predecessor, Prof. Dr. Albert Leemann, for their valuable supervision and support. Furthermore, I would also like to thank Prof. Dr. Harold Haefner for his support.

This study was connected to the Indo-Swiss research project "Towards Sustainable Development: An Actor-Oriented Perspective." This project and my research were financed by the Swiss National Science Foundation under the Priority Programme "Environment" (Module 7, project-no. 5001-038166.93). The project was carried out jointly by the Centre for Development Studies (CDS), Trivandrum, Kerala, and the Study Group on Institutions, Human Action and Resource Management, Department of Geography, University of Zurich. I am grateful to Samuel Wälty of the Study Group, for internal guidance during the project phase and for advise on the thesis. I would also like to thank the other project-team members: Prof. Dr. K.N. Nair (also for his comments on the thesis), Prof. Dr. P.S. George, Urs Geiser, Govindaru V., Jom Jacob, N.C. Narayanan, Antonyto Paul, Andreas Tarnutzer and, especially, Marco Pronk, for his valuable suggestions, which improved both field research and book manuscript, as well as for his companionship.

The Ph.D. thesis was reworked during my affiliation with the School of Oriental and African Studies, University of London. My thanks are to Robert Bradnock, Head of the Geography Department, for his interest in my research and support of this publication. The post-doctoral research was financed with a research stipend of the Swiss National Science Foundation.

I am also grateful to Ragesh Cheemara and Syed Ibrahim, for their excellent interpretation of interviews and insights into the Keralan way of life, as well as for their friendship. Furthermore, I would like to thank Joan Campbell for stylistic editing of the book manuscript.

I am also grateful for the insights, comments and support of the following people: Amitabh, Norman Backhaus, Barbara Harriss-White, Joy, Michael Kollmair, Coonoor Kripalani-Thadani, Vijaya Kumari, Heidi Meyer, Govindan Parayil, Phil Roy, Emanuel Schmitt, Arvinder Singh, Sunil S., Surya, Philippe Van Kerm, Sebastian Wilson and, especially, Lori McDougall and my parents.

Glossary

Adivasi	indigenous people of India; neither Aryans nor Dravidians
beedi	type of cigarette (rolled tobaccco leaf)
bund	embankment
coir	fibre of coconut husk
copra	dried coconut meat
jenmi	landlord
kadu	light natural forest
kamandar	superior tenant
kayala	terracing slopes with mudwalls on which reeds are grown; traditional soil-conservation measure in Kerala's midland
Krishi Bhavan	local agricultural offices of Kerala's Department of Agriculture
kudikidappu	hutment dwelling on the landlord's land
kudikidappukar	landless person having the hutment dwelling on the landlord's land
kudivaruppu	subcontracting of home-cottage industries by big firms
latex	milky, white fluid of rubber trees; raw material for the production of natural rubber
Malayalam	Dravidian language; language of Kerala
Malayalees	people of Kerala
marumakkathayam	matrilineal system of inheritance; formerly widespread among the *Nairs*
Nair	a dominant middle caste in Kerala
Namboodiri	a high caste in Kerala
panchayat	local political-administrative unit in rural areas
panchayati raj	local self-government
princely state	territory under indirect British rule and governed by a local *raja* before India's independence in 1947
Pulaya	a low caste in Kerala (Untouchables)
raj	rule
raja	king
tharavadu	household of a matrilineal *Nair* clan
vanam	dense rainforest
verumpattamdar	cultivating tenant

1 Introduction

1.1 Environmental Degradation and Markets in Developing Countries

At least since the 1992 UN Conference in Rio, it has been widely known that environmental degradation is an increasingly serious problem for Third World countries, and one that has a major impact on the health and livelihoods of the world's poor. Yet despite that knowledge, public attention tends to focus on global, forward-looking concerns such as the greenhouse effect and loss of biodiversity instead of very real current problems like soil degradation, water contamination and air pollution.

Third World environmental degradation results from increasing industrialization and urbanization, as well as the extensification and intensification of agriculture. The most serious agriculture-related effects include tropical deforestation, overgrazing of rangelands, soil erosion, salinization, and water and soil contamination (Bifani 1992: 99). Pollution and depletion of bio-physical resources are critical concerns for the growing populations in developing countries, where agriculture is a major source of income, employment and food security. Environmental protection and human well-being are interrelated, as shown in the World Commission on Environment and Development report *Our Common Future* (WCED 1987). It is now necessary to promote sustainable development, reconciling socioeconomic development and environmental protection.

Scholars have explained agriculture-related environmental degradation in developing countries in terms of underdevelopment and poverty (WCED 1987); capitalism and imperialism (Redclift 1987; Gadgil & Guha 1992); economic backwardness and technological stagnation (Karshenas 1994); spread of Green Revolution technologies (Glaeser 1987); underpriced natural resources and ill-defined property rights (Pearce 1988); resource use beyond the carrying capacity in tropical regions (Kirchner *et al.* 1984) and population pressure (Ehrlich & Ehrlich 1990). Recently, in light of economic trends toward global integration, policy liberalization and increased agricultural trade, scholars have begun to pay more attention to the impact of the market on environment and development. However, the study of this relationship has often been simplistic and general in nature.

1

Environmentalists, for example, tend to see the spread of the market as both destructive and inequitable: profit-driven development needs to clear land and people in order to build cities, factories and hydroelectric dams, and the acceleration of that process encourages overexploitation of natural resources by the poor and overconsumption by the rich (Mies & Shiva 1993). Further, growing urban-industrial market demand promotes the cultivation of resource-depleting cash crops that displace technologically appropriate, subsistence-oriented cropping patterns and the general eco-logical prudence of local communities (Kothari & Parajuli 1993).

At the other extreme, neoliberal advocates of free trade regard the market as the saviour of development and a protector against environmental degradation. According to the logic of "free-market environmentalists," the extension of private property rights to environmental resources would automatically lead to economically and ecologically efficient resource allocation, which would in turn permit the "optimal" level of environmental protection. State involvement in this process would only lead to greater inefficiency, by hindering the voluntary exchange of environmental property rights between consenting parties (Anderson & Leal 1991), while subsidies of agro-chemicals and food would further lead to inefficient cropping patterns and unsustainable agricultural practices.

There is empirical evidence to support both these positions. But ideology has tended to encourage both sides to make selective use of examples and to ignore the diversity of environmental problems and socioeconomic processes at a local level. Even more complex theories of environment and development have generally failed to discuss the role of markets explicitly. The statement still seems to be true that "sustainable development is the objective of many perspectives on the environment, but the role of the market in defining the various outcomes is considered in few of them" (Redclift 1987: 11).

The aim of this book is to fill that gap, contributing to a more detailed understanding of the increasingly important relationship between crop markets, agricultural practice and sustainable development. Case studies on pineapple and cashew cultivation in selected localities in the South Indian state of Kerala are included to demonstrate the way in which these factors interrelate and how market-induced changes affect human well-being and environmental sustainability.

Empirical evidence can both challenge theoretical approaches to sustainable development and assist in the formulation of environmental policy, particularly by aiding the appraisal of various market-based instruments and regulatory measures. Although Kerala is surely a special case within India because of its excellent social indicators, the results of this study may nevertheless indicate a number of lessons for other Indian states and developing countries.

1.2 Key Concepts: Social Practice and Real Markets

This book maintains that the above-mentioned neoliberal-economist and structuralist-environmentalist positions work with disputable assumptions regarding cultivators. While neoliberals assume the farmer (as well as other agents) to be a "rational" *homo economicus* constrained only by scarce resources, structuralist approaches regard peasants as victims of circumstances, including socioeconomic conditions such as market forces.

But if one accepts the sociological perspective of Giddens (1979, 1984), human beings are not fully determined by structures, but are knowledgeable and capable agents. Therefore, knowledgeability implies that agents know a great deal about the conditions for – and consequences of – their actions, and are able to give reasons for their conduct. Capability means that all agents, through the consequences of their actions, exercise some – though not the same – degree of power that can transform existing states of affairs or courses of events. In other words, individuals are capable of doing things differently or even of making a difference.

Yet, the agent's consciousness, intentionality and rationality are not the main characteristics of human agency. For most actions, consciousness is not "discursive" but "practical," belonging to tacit stocks of knowledge. That is, only if they are asked, are the agents able to offer reasons for their conduct *(rationalization* of action). Moreover, human agency is neither merely voluntaristic nor reducible to subjective intentions. On the contrary, for the most part, human agency is institutionalized, structured and routinized (and therefore referable to as *social practice)*.

In this book, cultivators are assumed neither to be independent of economic, social, political, cultural and ecological conditions, nor to be completely determined by these conditions. Rather, I believe that cultivators as well as other social agents have the capability to partially change conditions; and conditions, in turn, structure human agency without determining it completely. For example, cultivators are capable of developing agricultural technology and, possibly, also of influencing agricultural relations and marketing. Paying special attention to social practices, this book places much stock in the cultivators' capability, knowledge, perceptions and intentions, and also the cultivators' own explanations for why they apply particular agricultural practices. This perspective also recognizes that knowledge, power, access to resources and perceptions of environmental change may not be homogenous, but differentiated by class, caste, ethnicity, gender and locality.

Another root of the sweeping generalization of the neoliberal and environmentalist positions is their use of an abstract concept of "the" market or market "forces" that neglects actual processes of buying and selling, as well as qualitative differences and social embeddedness of markets. However, empirical studies have found that markets are unique in space and time. Particularly in developing countries, markets tend not to be fully developed and they operate under conditions of imperfect competition. Furthermore, markets are embedded in social and political structures.

As opposed to the abstract markets, these markets are referred to as "real" markets (Mackintosh 1990), "actually existing" markets (Hewitt de Alcántara 1993) or "unromantic" markets (Harriss-White 1996b). *Real markets* – in the following simply referred to as markets – are neither synonymous with the private sector nor compatible with the neoliberal idea of the "free" market. Rather, all real markets are regulated to a certain degree – shaped formally and informally, by the state or other national, regional and local social institutions.

As markets are embedded, crop- and locality-specific, the crucial question is not whether more or less market (as opposed to regulation) is needed for development but rather what *kind* of markets or market regulation is needed; and what type of interrelation between markets, the state, social structures, technology and infrastructure contribute either to sustainable development or to underdevelopment and environmental degradation.

1.3 Chapter Overview

Following this outline of basic concepts, Chapter Two leads to the issue of environment and development. As indicated, environmental change is interpreted with the concept of social practice: Socioeconomic and environmental conditions influence agricultural practices, which, in turn, have consequences for the environment. After presenting the mainstream concept and constituents of sustainable development, the chapter discusses complementary and conflicting theories of environmental change and development (environmental economics, political ecology and community-based development). Special attention is given to the significance of crop markets for sustainable or unsustainable development.

Chapter Three focuses on market exchange, agricultural decision-making and socioeconomic development. Agricultural decision-making and overemphasis on farmers' profit-maximizing motives are put into perspective with the importance of routinized social practice, and of security and non-economic goals. Then, the chapter recapitulates the concept of real markets and examines neoclassical, neo-Marxist and new institutional development theories in order to better understand the relationships between agricultural markets, state interventions, agrarian relations, infrastructure, technology and agricultural development. Environmental and development theories are then synthesized into three viewpoints that contain competing hypotheses about the relations between markets, environment and development. In this book, these viewpoints are used as "sensitizing devices" for the collection, analysis and interpretation of the empirical data.

Chapter Four outlines the methodology of the empirical study, clarifies the relation between the applied macro- and micro-level research, and introduces the two case studies. This chapter categorizes the conditions for agricultural practice into various technical-material and socioeconomic factors (including crop markets), operationalizes the concept of sustainable development with regard to socioeconomic and environmental consequences, and develops the cultivator classification used in the case studies. While the part on Kerala makes use of secondary material, the case studies rely to a great extent on primary data collected with qualitative methods over 15 months in 1994 and 1995. Starting with the cultivators' practices as well as with their own views regarding pineapple and cashew cultivation, the case studies analyze the relative significance of crop markets for agricultural processes as well as the implications for sustainable development.

Chapter Five presents the regional context of Kerala. Apart from general information on Kerala's history, geography, society, environment and development, this chapter discusses the general conditions for farming in Kerala. Striking features of Kerala's agricultural sector include effective implementation of land reforms, small-holding structure, predominance of part-time farming, formalized labor relations and comparatively high wage rates, well-developed product markets and well-developed formal credit markets. An overall impression of the relative importance of particular socioeconomic and technical-material factors for overall agricultural development and environmental change in Kerala is also provided.

Chapter Six and Seven present the case studies on pineapple cultivation and on cashew cultivation, respectively. They first give background information on general patterns of production and trade, as well as descriptions of pineapple cultivation in Vazhakulam, Ernakulam District (Central Kerala), and of cashew cultivation in Mattanur-Iritty, Kannur District (North Kerala). Then, the agricultural, socioeconomic and technical-material processes related to pineapple cultivation and cashew cultivation, respectively, are analyzed. Each chapter starts with the cultivators' own views. This is followed by an analysis of the significance of crop markets, agrarian relations, technology, infrastructure, bio-physical factors, cultural values and motives of individuals. Finally, the impact of market-induced agricultural processes on sustainable development is assessed.

Chapter Eight summarizes the main findings of this study. I will argue that crop markets can work both for and against sustainable development, and that their effect depends greatly on the socioeconomic and technical-material context in which they are embedded. Under conditions of relatively equitable sociopolitical structures, markets support development, and gains from trade are likely to be distributed over various sections of society. On the other hand, market-induced growth deepens negative effects created by other factors such as unequal access to resources or application of inappropriate technologies. Commercialization tends to lead to agricultural specialization, which can either involve environmentally unsustainable intensification or make possible concentrated cultivation of cash crops in areas where this is ecologically most suitable. This chapter also spells out the implications of these findings for development theory and government policy, and offers suggestions regarding future research directions. The book suggests that appropriate regulation of markets and consumer pressure – rather than either "free" trade or withdrawal from markets – have the potential to direct agricultural producers to apply more sustainable practices.

2 Environment and Development

The notion of sustainable development has drawn attention to environmental issues of development. As a consequence, a variety of explanations of development and environmental change have evolved since the 1980s such as environmental economics, political ecology and community-based development. These approaches to sustainable development refer to markets as a cause of, or potential solution for, environmental degradation. They also have both shaped and contested the evolving mainstream of sustainable development. In order to discuss sustainable development, however, I first need to clarify the ontology of the relation between environment and society.

2.1 Society-Environment Relations

In today's world, environmental change is to a high degree human-made. Human activities, including agricultural practices, lead to environmental change and have implications for development. Furthermore, all human actions and their implications are situated in an economic and social context in which individual interests vary and may conflict. Consequently, environmental change is as much a *social* issue as a physical phenomenon. Therefore, it is useful to draw on the social sciences in order to elucidate the relationship between society and the environment, and to study the origins and social implications of environmental change.

2.1.1 Social Practice and Environmental Change

Historically, the study of the relationship between humans and the environment has been a domain of geography. In the late 19th and early 20th century, most geographical approaches involved an environmental determinism (geodeterminism) that saw humans and societies as a product of their material environment. By the 1920s, geodeterminism was challenged by the concepts of possibilism (French regional school of geography), human

7

ecology (Chicago School) and cultural landscape (Carl Sauer). In these approaches, however, the environment was regarded as something external to society and culture. In the 1950s and 1960s, the study of environmental relations faded on the academic agenda of human geography, as geography was redefined as an "objective" science that discusses spatial relations rather than society-environment relations (Barnes & Gregory 1997: 174-175). Moreover, in radical geography – the critical Marxist response to "objective spatialism" – spatial relations remained the focus of geographical theory (FitzSimmons 1997). Environmental relations were paid at least some attention in the theory of the social production of nature (see N. Smith 1984). More recently, the social *construction* of nature has been emphasized. This theory views the environment as inseparable from society. Knowledge about the environment is constructed in situative, culture-specific and also researcher-specific contexts leading to various *discourses* (webs of concepts, statements and practices) about the environment (Barnes & Gregory 1997: 178). Moreover, the definition of a particular situation as an environmental *problem* involves a judgement based on social, cultural and political orientation (Redclift & Benton 1994). Yet, there are different degrees to which the environment is socially constructed:

> [T]here are aspects of the environment that are more open to alternate interpretations and therefore can be said to be more socially constructed than others. This means that some are closer to "facts" and a single and unopposed interpretation from a variety of actors than others. (Blaikie 1995: 9-10)

In this study, social constructions of the environment and of environmental problems are acknowledged but the focus is on *social practices* that bring about environmental change. Social practices reproduce the social and environmental setting and are, at the same time, influenced by this setting. As a *condition* for human agency – a particularly important condition in the case of agriculture – the environment becomes an integral part of society. Furthermore, the environment and environmental change have social content and meaning – not only as social constructs but also as *consequences* of social practices. Consequently, environmental change has implications for human development.

2.1.2   Conditions and Consequences of Social Practices

My general research perspective, i.e., to relate society and the environment to each other through social practices, is based on elements of *structuration theory* (Giddens 1979, 1984) and of the *three-world scheme* (Schutz 1962).

Giddens' structuration theory is an attempt to overcome the dualism that has existed in the social sciences between structural-determinist and interpretative, voluntarist approaches to human agency. According to this sociological perspective, human beings are not fully determined by structures, as the structural-determinists suggested, but are knowledgeable and capable agents. Yet, intentionality and consciousness are not the main characteristics of human agency, as interpretative approaches suggested, but human agency is, for the most part, institutionalized and routinized social practice (see Ch. 1). Giddens has addressed the mutual dependence of human agency and social structure with the concept of the *duality of structure:*

> By the duality of structure I mean that the structural properties of social systems are both the medium and outcome of the practices that constitute those systems.... The identification of structure with constraint is also rejected: structure is both enabling and constraining.... (Giddens 1979: 69)

Individuals draw upon and are influenced (enabled and/or constrained) by social structure when they conduct actions. At the same time, actions reproduce (or even transform) this structure through intended and unintended consequences, which form new constraining and/or enabling conditions for future action. In this way, the concept of the duality of structure is implicated in the ramified meanings that the conditions and consequences of action have. Furthermore, structuration theory emphasizes regularities in social practices. Those practices with the greatest time-space extension (distanciation) within a social system are referred to as institutions, which are the basic building blocks of social systems. Social systems are not congruent with social structure, which only exists as an ordering principle. Social systems rather *have* particular structural properties.

Structuration theory primarily deals with social relations rather than with relations between society and the environment. However, particularly regarding agricultural practices (as long as agriculture relies on "natural" soils and atmospheric climate) environmental relations also need to be considered. A way how to look at environment-society relations is implied in the *three-world scheme* proposed by Schutz (1962, 1982) and presented in Werlen (1993). In Schutz's phenomenology, three ontologically different "worlds" are specified: the subjective, the social, and the physical. The subjective world contains the stock of the subject's knowledge that is, to a great extent, acquired through socialization. The social world is experienced primarily through interactions and in the form of other agents' typical expectations in typical situations. The physical world comprises material objects, including the agent's body. The agent constitutes the physical world via the experience of his or her own body in movement, relating it through the critical stock of knowledge. In action situations, the ontologically different social and physical worlds are brought into relation with the subjective world; the social and the physical (the society and the environment) become integrated (Werlen 1993: 78-79).

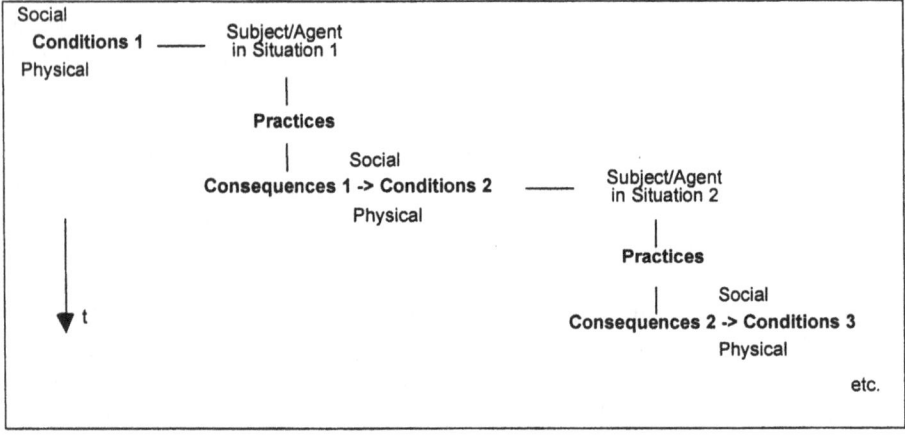

Fig. 1 Concept of Social Practice
Source: This conceptual representation draws on Werlen's model of social action (1993: 11-13) and on discussions within the CDS-GIUZ project "Towards Sustainable Development: An Actor-Oriented Perspective," in particular with M. Pronk.

Fig. 1 shows that the agent, being in a particular situation, projects actions according to his or her orientation, intention, motives and knowledge (subjective component of human agency). Furthermore, the agent takes social and physical conditions into consideration, both as enabling, and constraining, the intended practice. Therefore, the physical setting (as well as the social one) does not just "environ" the subject but is integrated into human agency. However, the agent does not acknowledge all the relevant social and physical conditions. Unacknowledged conditions often result in *unintended* consequences. Another type of unintended consequences may also appear: negative social or environmental consequences that agents may anticipate but nevertheless accept in view of the main intention. Furthermore, intended and unintended consequences alter the social and physical setting and form a new situation for the agent and often for other agents, too.

An agent may not have the same intentions for every action, nor approach each particular action with an equal degree of consciousness and intentionality. One-time decisions, for example, are usually taken more consciously than repeated, routinized practices. Also, when conditions change, agents tend to conduct actions more consciously as they "respond" to new constraints or make use of new opportunities. In relatively stable situations, on the other hand, it is more likely that traditional practices will prevail. Furthermore, some actions of individuals are structured by the objectives, rules and resources of the organization to which the individual is related rather than by the individual's own motives. In such cases, not only the goals of the individual but also the objectives of the respective organization need to be considered.

The outlined perspective forms the general guideline for this book. For the study of environment and development, however, it is more appropriate to use concepts that are commonly used in the life-world and that are adjusted to the research topic rather than to ontological theory. Giddens himself has pointed out that:

> There is, of course, no obligation for anyone doing detailed empirical research, in a given localized setting, to take on board an array of abstract notions that would merely clutter up what could otherwise be described with economy and ordinary language. The concepts of structuration theory, as with any competing theoretical perspective, should for many research purposes be regarded as sensitizing devices, nothing more. (Giddens 1984: 326)

With regard to environment and development, *sustainable development* has become the latest development catchphrase in political discourse, geographical and other research.

2.2 Sustainable Development

In the 1960s and 1970s, environmentalists started to draw the attention to contradictions between development and environmental protection. This discourse has been widely replaced with the debate about how development and environment can be reconciled and how sustainable development can be achieved (Lélé 1991: 107). This new emphasis evolved in the 1980s – particularly with the publication of *Our Common Future* (also called the *Brundtland Report)* by the World Commission on Environment and Development (WCED 1987). The phrase "sustainable development" became further popularized with the Rio Earth Summit in June 1992 (United Nations Conference on Environment and Development).

Today, a broad range of non-governmental, governmental and multi-lateral development organizations, including the World Bank, adhere to the concept of sustainable development, which seems to have become the development paradigm of the 1990s. The political use of this concept by ideologically such different groups as transnational companies and environmentalists has resulted in conflicting interpretations, many of which are contrary to the initial idea of sustainable development. These include notions of "sustainable, unlimited growth" and, at the other extreme of the spectrum, "deep ecology," aiming at metaphysical harmony between humans and nature (Arts 1994: 6). Despite the many conflicting interpretation of sustainable development, a discernible *mainstream* has evolved (Lélé 1991; B. Adams 1993).

2.2.1 The Mainstream Concept

The Brundtland Report represents the mainstream concept of sustainable development best. The report also put forward a widely accepted definition of "sustainable development":

> Sustainable development is development that meets the needs of the present without compromising the ability of future generations to meet their own needs. (WCED 1987: 43)

In addition to this general definition, the report suggests more elaborate operational objectives, such as reviving growth; changing the quality of growth; meeting essential needs for jobs, food, energy, water and sanitation; ensuring a sustainable level of population; conserving and enhancing the resource base; reorienting technology and managing risk; merging environment and economics in decision-making; and reorienting international economic relations (WCED 1987: 49). The strategy for achieving these objectives requires an economic system that generates surplus and preserves the ecological base; a technological system that provides for new solutions; an international system that fosters fair patterns of trade; a social system that is able to solve tensions; a flexible administrative system that has the capacity for self-correction; and a political system that secures citizen participation in decision-making (WCED 1987: 65).

As a consequence of this broad range of objectives and strategies, the concept of sustainable development has the potential to build a broad consensus among adherents of competing development theories, as well as between "developmentalists" and environmentalists:

> The current state of scientific knowledge ... about natural and social phenomena and their interactions leads inexorably to the conclusion that anyone driven by *either* long-term self interest, *or* concern for poverty, *or* concern for intergenerational equity should be willing to support the operational objectives of SD [sustainable development]. (Lélé 1991: 612)

The notion of sustainable development has transformed the environment-development debate fundamentally. As opposed to the ecocentric and neo-Malthusian environmentalists who dominated the discourse in the 1970s, the proponents of sustainable development reject the assumption that economic growth inevitably leads to environmental degradation or, in turn, that environmental constraints imply narrow, fixed *limits to growth* (Meadows *et al.* 1972). On the contrary, sustainable development suggested that many environmental problems might actually originate from the *lack* of development (i.e., that poverty might be a primary cause of environmental degradation) and that environmental degradation, in turn, can reinforce poverty (i.e., the poor, whose livelihoods are often directly dependent on natural resources, might be hit most severely by environmental degradation). The interrelation between poverty and environmental degradation is characterized in this concept by complex interrelations that also include determinants such as access to resources, technology and cultural values (see Fig. 2).

In practice, mainstream sustainable development emphasized technical-economic factors – such as inadequate technology, deficient managerial capability, undefined property rights, wrong pricing and subsidy policies – as main causes of environmental degradation; unequal access and cultural values were considered only marginally. This meant that the suggested solutions were also essentially technical-economic (e.g., technological reorientation rendering industrial and agricultural production less harmful, less resource-intensive and more productive; economic policy change that incorporates environmental considerations; clear definition of property rights). While the incorporation of non-governmental organizations and of local participation was promoted, calls for more radical sociopolitical reforms (e.g., land reform) or changes in cultural values (e.g., away from materialistic, consumption-oriented lifestyles), by and large, remained rhetoric (Lélé 1991: 613). In other words, the mainstream concept of sustainable development was *reformist* rather than radical (W.M. Adams 1990). Because it rejected ideas of ecocentrism and fixed environmental limits to development, sustainable development has also been said to represent a *renewal of modernism* (Redclift 1994).

Furthermore, the all-inclusive concept has not only led to the popularity of sustainable development, but also to its proneness to misinterpretation, and its use as a "container term" (Arts 1994) and political "slogan" (Adams 1993). Instead of dropping the term (and possibly replacing it with another slogan), Arts (1994) argued that it rather needs to be clarified in theoretical and practical terms, and distinguished from interpretations that are impracticable or contrary to the initial idea of sustainable development.

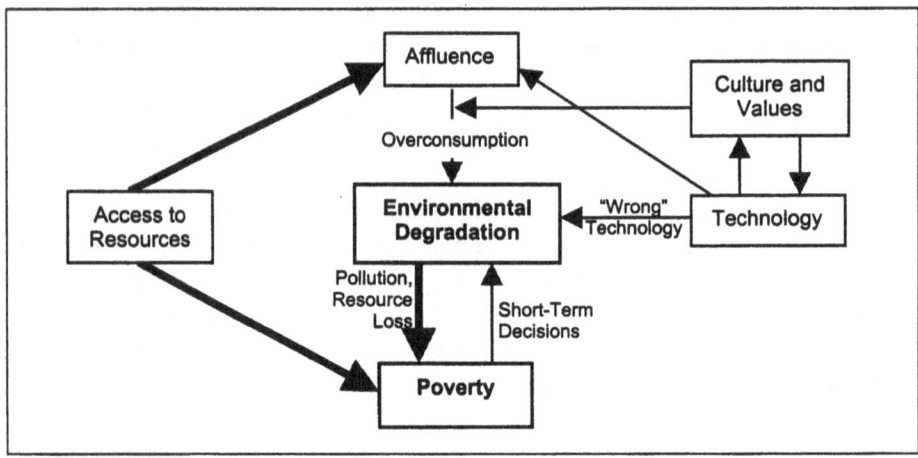

Fig. 2 Relation Between Poverty and Environmental Degradation
Source: Lélé (1991: 614).

2.2.2 Components of Sustainable Development

Sustainable development consists of two elements: *development* and *sustainability*. Development can be defined very broadly as the improvement of the living situation of all humans. Development comprises economic aspects (e.g., economic growth, poverty alleviation, employment), social aspects (e.g., basic needs, health, education, equality, social justice) and political aspects (e.g., participation, democracy, peace), all of which are interrelated and are both means and ends of development.

The notion of sustainability has incorporated a long-term view and consideration of environmental issues into the development discourse. *Sustainable* means both tolerable and being able to keeping on something more or less indefinitely (Arts 1994: 8-9). More specifically, *sustainable development* refers to the continued improvement of human well-being that has tolerable externalities on the local, regional and global environment, other social groups and future generations. In my view, development is to be sustained and not economic growth, agriculture, a particular land use or the environment *per se*. Rather, the mentioned elements contribute to sustainable or unsustainable development. In this book, a human-centered view of sustainable development is suggested, one that shares basic elements with the concept put forward by Arts (1994). Thereby, it is most useful to look at three components of sustainable development (environmental, economic and social) in their interrelation.

The environmental component includes maintenance and enhancement of essential ecological processes, biological diversity, and the natural resource base. Environments have the properties of sensitivity – degree to which human practices or broader bio-physical changes affect the environment – and of resilience – the ability of the environment to absorb change and respond positively to land management (Gibbon *et al.* 1995: 33). Environmental sustainability is important for human development because we humans are, through our bodies, part of nature. Thus, physical conditions are the base of our survival and social and economic life. This view of nature in relation to humans can be called *anthropocentric*, as opposed to *ecocentric*, a view that ascribes intrinsic value to ecosystems, fauna and flora. Loss of biodiversity, for example, is regarded as detrimental because "genetic erosion" may obstruct biological evolution and ecological stability, both of which are important for human survival and development in the longer term.

The economic component of sustainable development comprehends economic growth that satisfies the needs of the present generation and that retains the resource potential to support future generations. Economic growth *per se* is neither necessarily opposed to nor favorable to environmental sustainability. The outcome depends on the nature of economic growth; that is, to what degree economic growth implies the depletion of non-renewable resources and the overuse of renewable resources including the sink capacity (the regenerative capacity of the environment to absorb waste). To some extent, the nature of economic growth can be made more environmentally friendly and resource efficient through the development of appropriate technologies and substitutes for non-renewable resources. In developing countries, economic growth is important to provide the means to meet basic needs, to alleviate poverty and to generate employment. However, growing GNP per capita does not automatically lead to development. For development, economic growth and access to resources must be distributed over all sections of society. If, on the contrary, high economic, social and political inequalities persist, *growth without development* is likely to occur. So, growth is a necessary, but not a sufficient, condition for sustainable social development. On the other hand, aspects of social development such as human health and education are important, yet not sufficient, for sustainable economic development.

The social component of sustainable development goes beyond improvement of basic needs, human health and education to include social harmony within and between societies. Social harmony usually requires social justice, low economic inequality and some degree of political equality, including participation and empowerment of weaker sections of the society. High local and global inequality, on the other hand, can lead to the unsustainable use of resources, both by the affluent (over-consumption) and the resource-poor (degradation of the environment because of poverty and limited access to resources). Moreover, inequality of resources can even become a reason for war, which is probably the most unsustainable situation thinkable – environmentally, economically and socially. For social sustainability, the maintenance of essential government support systems for the poor and community safety networks can also be important. However, government services cannot be sustained in the long run when economic development does not take place and the state lacks the necessary financial resources. In the light of generally reduced public spending, many non-governmental organizations try to strengthen or build up community networks. Furthermore, these organizations seek to incorporate popular participation into the development process.

This human-centered, integrative concept of sustainable development is normative rather than explanatory. Also, the analytical use of the concept can be very complicated. In practice, spatial and temporal externalities may be hidden; and trade-offs between socioeconomic development and environmental sustainability are difficult to weigh against each other. Because demographic trends and technological progress are unpredictable, the needs of future generations remain an unknown dimension. Measuring sustainability also involves methodological problems of collecting, selecting, quantifying and comparing environmental indicators. Furthermore, sustainable development depends on the capability of different social groups to cope with environmental change. For these reasons, sustainable development requires operationalization to become useful for empirical research (see Ch. 4).

After having discussed about what sustainable development ought to be, the following section introduces approaches to sustainable development that include theoretical analyses of the causes of unsustainable development and practical solutions to solve environmental problems.

2.2.3 Approaches to Sustainable Development

The mainstream concept of sustainable development can be interpreted as a mixture of elements of particular theories of environment and development. Conversely, some of these theories, which tend to be more coherent than the mainstream concept, may be regarded as variations that emphasize different operational objectives of sustainable development. Competing explanations of underdevelopment and environmental degradation have also proposed a range of solutions to overcome underdevelopment and environmental problems, and to achieve more sustainable development. Still other theories of environmental degradation, however, stand in opposition to the mainstream concept. They are either radical, ecocentric, anti-modernist, or post-modernist. Table 1 contains short descriptions of theories of development and environmental change, focusing on their definition of the problem, their explanations for the current difficulties, and their policy suggestions, and displays a possible, though incomplete, classification of these theories into seven "ideal types."

The theories incompatible with the normative concept of sustainable development (utopian/radical ecology, neo-Malthusian environmentalism, deep ecology/ecofeminism, cornucopian technocentrism) either make simplistic generalizations about the relationships between markets, environment and development (see Section 1.1) or do not discuss them explicitly. Therefore, they are only of limited utility for this study. The following sections only discuss arguments of environmental economics, political ecology and community-based development. These theories give explanations of underdevelopment and environmental degradation and contain particular hypotheses about the relation of markets and environmental change.

Table 1   Theories of Development and Environmental Change

Variations of Mainstream Sustainable Development

Environmental Economics

Important References: Pearce 1988; Pearce *et al.* 1989; Turner *et al.* 1994

Theoretical Heritage: Neoclassical economics

Highlighted Aspects of Mainstream Concept: Economic sustainability; growth; merging environment and economics in decision-making

Definition of Problem: Underdevelopment as reduction of total capital stock; environmental degradation as reduction of natural capital stock

Causes: Misallocation of resources due to market distortions (underpriced natural resources; unpriced environmental services)

Solution: Removal of market distortions (e.g., cuts in subsidies); creation of markets for the environment (environmental tax, tradeable resource-use certificates); definition and enforcement of clear property rights

---

Political Ecology

Important References: O'Connor 1988; Blaikie 1985; Redclift 1987, Bryant & Bailey 1997

Theoretical Heritage: Marxist-influenced political economy, cultural ecology, radical development geography

Highlighted Aspects of Mainstream Concept: Social sustainability; unequal access to resources; international economic relations

Definition of Problem: Underdevelopment as exploitation of humans; environmental degradation as exploitation of nature

Causes: Maldistribution of resources (on international, national and local levels) under capitalism; extraction of surplus from the poor, who are consequently forced to overexploit bio-physical environment; overconsumption by the rich

Solutions: Environment-livelihood struggle of grassroots organizations; state interventions

---

Community-Based Development

Important References: Chambers *et al.* 1989; Ghai & Vivian 1992

Theoretical Heritage: Indigenous knowledge; agrarian populism; (development practice)

Highlighted Aspects of Mainstream Concept: Social sustainability; appropriate technology; participation

Definition of Problem: Underdevelopment as failure to meet basic needs; environmental degradation as locally unsustainable resource use

Causes: Inappropriate technologies, transfer of locally unsuitable "modern" technology; (import of) inappropriate institutions

Solutions: Participation; acknowledgement of indigenous knowledge, indigenous institutions and community organizations; comanagement

---

Table 1 (cont.)   Theories of Development and Environmental Change

---

Incompatible With Concept of Sustainable Development

Utopian/Radical Ecology

Important References: Zimmermann 1994

Theoretical Heritage: Utopian socialism; social anarchism

Aspects that Contradicts Mainstream Concept: Radical; not pragmatic

Definition of Problem: Environmental crisis as pillage of nature

Causes: Authoritarian social structures in capitalism as well as in state socialism

Solutions: Development of small-scale, egalitarian and anarchistic communities

---

Neo-Malthusian Environmentalism

Important References: Ehrlich & Ehrlich 1990; Meadows *et al.* 1972

Theoretical Heritage: Neo-Malthusian economics; systems ecology

Aspects that Contradicts Mainstream Concept: Assumption of necessary contradiction between environment and development; no consideration of social sustainability

Definition of Problem: Definite ecological limits of development; man-made sub-systems (economy and society) already beyond limits given by ecological system

Causes: Population growth; resource constraints; ecological constraints

Solutions: Top-down environmental management; restrictive environmental laws

---

Deep Ecology/Ecofeminism

Important References: Sessions 1995; Sachs 1993; Mies & Shiva 1993

Theoretical Heritage: Ecocentrism; post-modernism; feminism

Aspects that Contradicts Mainstream Concept: Ecocentrist; post-/anti-modernist; anti-development; no consideration of economic sustainability

Definition of Problem: Ecological crisis as abuse of nature

Causes: Anthropocentrism; Modernistic development project (incl. western science); logic of patriarchal domination that subjugates femaleness, emotion, body and nature

Solutions: Fundamental cultural change; ascribing intrinsic value to both nature and humans; dismantling patriarchy

---

Cornucopian Technocentrism

Important References: Bolch & Lyons 1993; Bailey 1993

Theoretical Heritage: Laissez-faire liberalism

Aspects that Contradicts Mainstream Concept: No consideration of environmental sustainability

Definition of Problem: – (denial of seriousness of environmental problems)

Causes: (Environmental crisis made up by environmentalists)

Solutions: Business as usual; technological innovations

2.3   Market Failure and State Failure

The failures of markets and governments to protect the environment have been highlighted by (neoclassical) *environmental economics*. As free-market environmentalists, environmental economists have supported suggestions for "getting the prices right." However, they have gone far beyond the neoliberal laissez-faire approach to prescribe market-based policy interventions in combination with direct controls and state regulations in order to achieve a more sustainable and efficient use of the environment.

In developing countries, agricultural inputs such as irrigation, chemical fertilizers and energy tend to be underpriced – usually as a consequence of sectoral government policies that are biased in favor of allegedly influential interest groups such as capitalist farmers. Such *state failure* and the resulting misallocation of resources does not only lead to economic inefficiency but also encourages wasteful use of agricultural inputs and the overuse of natural regions (Warford 1989: 15). Economies in which a large number of price distortions exist are often characterized by wasteful use of natural resources, often exceeding sustainable harvest levels. Therefore, price distortions can generate resource depletion (Pearce 1988: 112).

Furthermore, even prices for unsubsidized agricultural inputs may not reflect full costs. Usually, only production costs are considered (e.g., costs to produce chemical fertilizers). However, the industrial production of agricultural inputs also pollutes water and air, leading to *external costs* (that is, costs that are not borne by the producer but by the society as a whole, as well as by future generations; external costs include such things as depleted resource stocks and less productive, less healthy environments). Moreover, the farmers' application of particular agricultural inputs may generate additional external costs (e.g., pollution of water and soils through fertilizer use). Externalities such as river or reservoir sedimentation, soil erosion and desertification may render whole ecosystems less resilient to exogenous shocks such as climatic changes. For example, the effects of a "normal" monsoon may suddenly become catastrophic when degraded soils are not able to absorb the water anymore (Pearce 1988: 106).

In particular, externalities relate to unpriced environmental services such as the sink capacity (i.e., the capacity of the environment to absorb wastes). While the market system appears to use priced resources efficiently, it fails to bring about the efficient and sustainable use of unpriced environmental resources. This *market failure* (in the sense that no market develops rather than in the sense of what existing markets fail to achieve) is particularly common in the case of the allocation of *public goods*. (According to Turner *et al.* (1994: 77), public goods are characterized by non-exclusivity (i.e., people who are unwilling to pay can, in practice, not be excluded from the use of the goods) and/or by indivisibility (i.e., the goods cannot be subdivided in accordance with the price paid by particular users).)

Although neoclassical environmental economists may not have completely disregarded wider explanations of environmental degradation (i.e., unequal access to political and economic resources, population growth), they have focused on economic distortions and the resulting misallocation of resources. Consequently, environmental economists have only emphasized the economic incentives for more sustainable production patterns.

> [W]e are concerned with feasible, reasonably short-term incentive measures, rather than with wholesale changes in social processes that may take long periods of time.... [The latter] have to proceed, but they need not be exclusive of other, less complex and probably more immediately effective policy. (Pearce 1988: 112)

According to neoclassical environmental economics, counter-environmental price distortions regarding traded goods such as fertilizers, energy or irrigation water are to be removed in the first place. Presently non-traded goods such as environmental services, furthermore, should also be valued in monetary terms and integrated into benefit-cost analyses (i.e., external costs should be *internalized)*. Proposed instruments for the internalization of the environment into markets include environmental taxes on pollution and energy use, charges/bonuses aimed at improving environmental behavior of entrepreneurs, environmental bonds, tradeable emission permits and resource-use licences (see Doeleman 1992). Generally, these market-based instruments are guided by the *polluter-pays principle* that should encourage producers to apply cleaner technologies. Environmentally unsound production, on the other hand, would reflect in higher product prices. This may eventually discourage consumers to buy such products (Pearce *et al.* 1991: 8-9). Although these market-based instruments have been developed (at least in theory) for industrialized countries, environmental economists believe that these instruments would be the most useful and efficient means to protect the environment in developing countries too (Pearce *et al.* 1990).

Market-based instruments may need to be combined with regulatory instruments *(command-and-control* instruments) such as environmental standards, permissions and prohibitions. In developing countries, however, the prevailing institutional weakness of the bureaucracy may hinder effective implementation and supervision of market-based instruments and, even more so, of regulatory measures. According to environmental economists, therefore, institutions need to be strengthened. However, many environmental economists have failed to recognize the real problem: bureaucratic institutions in developing countries may not be weak but rather *inappropriate*. Alternative concepts have emphasized community-based strategies for environmental protection (see Section 2.5).

Definition and enforcement of clear property rights have a high priority on the agenda of environmental economists because without clearly defined, secure and enforced property rights, resource users have no incentive to protect natural resources and will neither respond to price incentives (Turner *et al.* 1994: 317). By contrast, an owner-farmer (or a long-term tenant) has a strong self-interest to utilize his or her resources in a sustainable way. In many developing countries, however, markets and clearly defined private-property rights for renewable resources such as arable land, forests or fisheries are not fully developed. Yet, also common-property regimes may guarantee environmental protection as long as community-based regulations do not break down (Pearce 1988: 108).

In summary, environmental economists have sought to incorporate the environment into the economic system by attaching monetary value to environmental conservation or degradation. However, they admit that environmental costs and the "demand" for environmental services are difficult to quantify. Critics have stressed that the monetization of the environment does not even make sense because people do not value the environment in monetary terms. Monetary values are always exchange values and therefore do not apply to the environment, which is a public good rather than a commodity traded in the market. Consequently, monetary valuations do not capture the worth of the environment to different groups of people (Redclift 1994: 28; Jacobs 1994).

Another set of criticisms has concerned the suggested market-based policy instruments. Many policy recommendations are logically derived from models based on *assumptions* such as "rational" individual behavior and perfect-market competition. Consequently, these recommendations are not value neutral. Indeed, by monetizing the environment, we place values on nature that reflect the priorities of our social system rather than the value of nature itself (Redclift 1994: 28). On the other hand, adopting more realistic assumptions (e.g., asymmetric information, time constraints, importance of cultural values for human agency) would indicate that agents may not respond automatically to economic incentives to protect the environment (Jacobs 1994: 82-83).

Even if agents such as cultivators responded "rationally" to price incentives and if agricultural inputs had the "right" price reflecting external costs, positive environmental effects could not necessarily be expected in every region. Reduced use of chemical fertilizers as a consequence of increased fertilizer prices, for example, can be as harmful as over-fertilization and can lead to exhaustion of the soil. Moreover, the application of a particular chemical may be harmful for a particular soil type but ecologically sound for another soil type. Generally, the pricing of the environmental impact of agricultural inputs faces major practical difficulties because the environmental impact is not proportional to the quantity used and, furthermore, depends on the peculiarities of the ecosystem as well as on the time at which the input is applied. Unlike neoclassical environmental economics, *ecological economics* has considered intrinsic characteristics of ecological systems in the attempt of integrating economics and ecology (see Costanza 1991; van den Bergh 1996).

Finally, even if an economically and ecologically "optimal" resource allocation could be achieved with market-based instruments, aspects of distribution and social justice would remain unconsidered (Martell 1994: 70). Just as food markets do not respond to human *needs* but to human *demand*, which is dependent on purchasing power (see A. Sen 1981), a market for the environment does not react to the needs of the poor (or to environmental "needs") but rather to the preferences of those with more purchasing power.

These criticisms show the limits of market-based policy instruments to achieve sustainable development. Yet I believe that such factors as under-priced resources, external costs and unclearly defined property rights deserve our attention as possible conditions for cultivators' social practices that lead to environmental degradation.

2.4 Capitalism, Inequality and Poverty

The environmental impact of exploitative capitalist structures, which lead to inequality and poverty in developing countries, has been emphasized by the Marxist-influenced approach of *political ecology*. Unlike environmental economists, political ecologists focused neither on problems of "optimal" resource allocation nor on policy instruments that are effective in the short term, but rather on "big" causes of environmental degradation, long-term social change and the actual effects that environmental degradation has on different social classes. As opposed to *scientific* ecology that remains in the realm of natural science, *political* ecology can be understood as an attempt to consider socioeconomic and political processes as explanations of environmental change (Little & Horowitz 1987). Political ecologists have also considered the social construction of the environment and, in particular, the ways in which ideas and discourses about environmental degradation and sustainable development are developed by different groups to facilitate or block a specific interest (Bryant & Bailey 1997: 21-22).

A common ground of different political-ecology approaches is to treat the environment as a social category. This perspective goes back to Marx's observation that production is a double nexus in which the relationship between humans and their environment is bound up with relationships among humans themselves (Collins 1992: 179). In the orthodox Marxist view or socialist-ecology perspective, the current ecological crisis has been interpreted as the *second contradiction of capitalism*. This contradiction refers to the inability of the capitalist production system to reproduce its own basic conditions (i.e., bio-physical conditions, human labor, communal conditions of production). Instead, capitalism undermines its own conditions (O'Connor 1988; Gorz 1980).

Yet, most political-ecology studies have used less rigid theoretical concepts that depart from orthodox Marxism. Blaikie & Brookfield (1987: 17) have given a working definition of political ecology:

> The phrase "political ecology" combines the concerns of ecology and a broadly defined political economy. Together this encompasses the constantly shifting dialectic between society and land-based resources, and also within classes and groups within society itself.

Political ecology has usually combined an analysis of localized, "place-based" environmental processes and social interactions with that of "non-place-based" factors and causes of environmental change such as the global economy, the state and the wider civil society. A *chain of explanation* may start at the site of environmental change, with the practices responsible for, and the specific agents behind, this change. However, since the causes of environmental degradation may be outside the afflicted area altogether, an analysis of the political economy is also needed (Blaikie 1995: 17-18). Political ecologists have emphasized that global processes of capital accumulation are linked to inequality and poverty at the local level; which, in turn, are implicated in environmental degradation (Collins 1992: 179). Despite the general attention to global capitalism, the many political-ecology studies have identified distinct, local-specific causes of actual environmental degradation in developing countries. The value of political ecology, therefore, is the investigation of diversity and heterogeneity of actual processes rather than the development of a generic theoretical explanation of environmental degradation.

Nevertheless, some general factors have commonly been mentioned as causes of environmental degradation. These include:
- unequal access to resources and unequal land distribution;
- poverty and exploitation of the poor through capitalist relations of production;
- pressure for subsistence farmers to cultivate cash crops;
- displacement of peasants into ecologically sensitive regions as a consequence of the spread of commercial farming in fertile regions;
- pro-rich state policies;
- large-scale development projects;
- activities of transnational companies;
- imbalance of international trade.

Colonialism and capitalism are regarded as having undermined pre-colonial relations of production as well as pre-colonial resource-management systems, which had been sustainable in most cases. Today, structural linkages between industrialized countries and developing countries continue to affect the environment.

> The penetration of the South by new agricultural technologies, marketing and contract farming, have also served to shift agriculture ... away from traditional, environmentally sustainable systems towards greater specialization and economic dependency. (Redclift 1987: 12)

The concept of the *simple reproduction squeeze* (see Section 3.4) has provided a common theoretical platform for the analysis of the contextual sources of localized environmental degradation. Faced with declining terms of trade, farm households have the choice either to reduce consumption, with negative effects on nutrition and health, or to intensify production, which may have ecological costs unless soil-replenishing measures can be afforded and are applied. A further alternative is to engage in off-farm employment and to migrate periodically. However, this may lead to a seasonal scarcity of labor in the agricultural sector that can also have negative ecological consequences. As a consequence of labor scarcity, labor-intensive but environmentally sound mixed-crop farming may be abandoned, or the maintenance of terraces, embankments, etc. may break down. Generally, indebted households enter these destructive dynamics with greater frequency. In particular, repayment schedules can force poor farmers to produce for short-term gain rather than for long-term sustainability (Collins 1992: 182-185). *Forced commerce* (see Section 3.4), furthermore, enhances extraction of surplus value from peasants and, therefore, may lead to over-exploitation of bio-physical resources, and to environmental degradation.

Moreover, because Third-World peasants regard the environment primarily as a source of their livelihood (rather than as an amenity or aesthetic asset, as many First-World environmentalists would have it), they may struggle to protect the environment and to get access rights to critical resources. Conflicts over access to resources and the socioeconomic, as well as political, ramifications of environmental change have become core topics for political ecology (Bryant 1992). Furthermore, increased attention to Third-World environment-livelihood movements, civil organizations, indigenous institutions and indigenous technical knowledge has moved Marxist-influenced political ecology closer to community-based approaches (see Section 2.5). Instead of romanticizing agrarian populists, however, political ecologists have continued to stress the importance of the wider political economy in which cultivators in developing countries are situated:

> When peasants and tribals are caught up in markets – and particularly in the contradictory demand resulting from simultaneous participation in several labor and commodity markets – their production practices become intimately linked to cycles of the world economy. The tendency of many recent environ-mentalist writings to use third world peasantries as examples of "organic communities" ... and to glorify their ecological practices once again neglects the real economic circumstances in which they must provision and reproduce themselves. (Collins 1992: 184-185)

With respect to policy instruments for sustainable development, political ecology is more radical but less concrete than neoclassical environmental economics. Sustainable development is regarded as "not just a pursuit of ecological guidelines and new planning structures, but an attempt to redirect change to maintain or enhance the power of the poor to survive without hindrance and to direct their own lives" (W.M. Adams 1990: 202). For orthodox Marxist development strategists, socialist revolution and extensive state intervention (e.g., land reform, state control over marketing) still formed top priorities. Because of mixed experiences with state interven-tions, however, more recent approaches have emphasized the role of social movements and community-based organizations for development. For the solution of environmental problems, planned intervention and democratic institutions that reflect the will of citizens rather than of consumers, and, especially, environment-livelihood struggles of community-based organizations are regarded as strategies capable of bringing about the social changes necessary for sustainable development. Markets, on the other hand, are still distrusted because of their exploitative structure. In my opinion, however, it is questionable whether states and community organizations are less likely than markets to fail in making development more sustainable. Even if institutions are democratic, they reflect the socioeconomic priorities of the present generation, but not those of future generations.

Stressing the impact of global capitalism on environmental degradation, the political ecologists of the early 1980s inherited Marxist structural determinism and economic reductionism, which are incompatible with the concept of social practice. However, the more recent political-ecology perspective, which pays more attention to human agency and interaction of various agents and organizations at the local, national and international level, offers useful insights on the relationship between markets and sustainable development; i.e., that social relations of production, inequality and poverty can be critical conditions for unsustainable social practices.

2.5   Capacity of Local Communities

As an alternative approach to both neoclassical and Marxist explanations of environmental change, *community-based development* has emphasized the local communities' capacity and the potential of participation to ensure sustainable development. Community-based development, which is a practical development strategy rather than a concise theory, has drawn our attention to the value of indigenous technologies and their particular relevance to environmental issues. This partially populist perspective has argued that "imported" technology, organizations and institutions are often unsuitable for sustainable resource use at the local level. Therefore, community-based development emphasizes the need for acknowledging indigenous knowledge, indigenous institutions and community participation.

Indigenous knowledge usually refers to knowledge that has been generated locally and that is unique to a community or society as opposed to knowledge that is generated through the global network of universities and research institutes. Yet, indigenous knowledge includes knowledge that had been generated elsewhere but has been transformed locally. Generally, it involves technology but also socioeconomic and cultural aspects.

Indigenous-knowledge studies have challenged the assumption of models of agricultural innovation (see Hayami & Ruttan 1985) that only "scientific" research stations can develop agricultural technology. Indigenous-knowledge approaches, furthermore, have rejected the static view and negative valuation of modernization theory regarding tradition, traditional technology and agriculture (see Schultz 1964). On the contrary, these approaches have stressed the great potential of indigenous knowledge for development in general and for environmentally sound resource use in particular. Indigenous knowledge systems have thus gained increased attention from development organizations and academics (e.g., Brokensha *et al.* 1980; Richards 1985; Chambers *et al.* 1989; de Boef *et al.* 1993).

Development experts have ignored farmers' experiments and innovations for a long time. A reason for this neglect was the scientists' unfamiliarity with the methods by which indigenous knowledge is stored, transformed and communicated. Obviously, farmers' records are not kept in written form and, therefore, are not easily accessible to scientists. Furthermore, the modernization approach (including the Green Revolution) trusted almost blindly in the benefits of modern science and the transfer of modern technology packages (Rhoades 1989: 4-5). However, farmers often do not adopt the "advanced" modern technology. Modernization theorists traced back non-adoption of technology packages to the poor farmers' economic constraints, conservatism and irrationality.

By contrast, indigenous-knowledge approaches have identified as the problem the transferred technology itself. Given the farmers' priorities, interests and local conditions, transferred modern technology packages are often inappropriate (Chambers *et al.* 1989: xix). Furthermore, indigenous-knowledge studies have shown that farmers adapt their knowledge to changed conditions and that they may also experiment with "imported" modern technologies. Far from being conservative, they prefer to *adapt* new technologies step-by-step and very selectively to their own farming and household conditions rather than *adopting* entire technology packages, which tend to be unsuitable for small-scale farming (Rhoades 1989: 5; Maurya 1989). Furthermore, farmers experiment with new environment-friendly technologies such as integrated pest management or green manuring (Santhakumar 1995). Moreover, archaeological and historical records have revealed that, even in the Neolithic period, farmers did not react passively to their environment but experimented actively with seeds and cultivation methods. In this way, over thousands of years, farmers in different areas of the world selected and domesticated all of today's major and minor food crops.

Since the 1980s, this indigenous knowledge has been increasingly considered as containing sophisticated, technical and organizational, forms of resource use and of agriculture that are specifically adapted to diverse local bio-physical conditions (de Boef *et al.* 1993: 1). Unlike previous romanticized views of past traditional societies, indigenous-knowledge approaches have become relevant to current development. Although, in some cases, traditional knowledge has lost its value and suitability under contemporary socioeconomic and demographic conditions, development projects still could learn from indigenous, use-oriented environmental know-how. On this basis, new concepts for an ecologically sound use of the natural environment could evolve. Indigenous land-use systems may form a local potential that should be developed rather than simply superseded by modern agricultural technology (Müller-Böker 1995: 378-379). Furthermore, well-functioning indigenous systems may also become models for other regions.

Whereas the "transfer-of-technology" approach is based on the knowledge of scientists in agricultural-research stations alone, community-based development starts with the farmers and their knowledge, problem identification, analysis and priorities. It then attempts to combine indigenous knowledge with "scientific" knowledge. While indigenous knowledge has the advantage of being adapted to the local bio-physical and socioeconomic conditions, "scientific" knowledge generally can adapt faster to changing conditions (Chambers *et al.* 1989). In this approach, participation clearly goes beyond the implementation of programs designed by "outsiders" to include participation in development planning.

Related to community-based development is an alternative view of the interaction of local people with their environment. Rural people in developing countries are regarded as neither necessarily negligent in terms of their resource management nor as invariably driven by poverty, population pressure and external forces to undertake environmentally unsustainable practices. Instead, cultivators are seen as capable, knowledgeable and concerned about their resource use and its environmental consequences. In many cases, moreover, rural people are proactive (as opposed to merely reactive) in *protecting* their environment, which naturally is valued as an important source of livelihood. To this end, collective action may be employed to resist unsound large-scale "development" projects (Ghai & Vivian 1992: 11-13). Furthermore, local resource use may be regulated using a whole system of indigenous institutions. These include rules about ownership, and access to natural resources and indigenous knowledge. In

this way, sustainable agricultural systems have often developed indigenously that (see Gibbon *et al.* 1995: 52-53):
- use renewable resources;
- do not rely on many external inputs;
- are diversified and structurally interlinked;
- focus on food security;
- are controlled through social sanctions and local policing.

As they have technical competence and a genuine self-interest in
protecting the environment, local communities seem effective to manage
and restore the environment. Particularly in developing countries, community participation is also believed to be the most effective strategy because
of market failures and state failures. Community-based development
therefore suggests that the implementation of sustainable development
should be based on local-level solutions derived from community initiatives
(Ghai & Vivian 1992: 1). Advocates of this strategy propose more
responsibilities for local communities, involvement of grassroots organizations and NGOs, administrative decentralization and *comanagement* (i.e.,
sharing of responsibility between the state and the local communities).
More conservative community-based approaches dispute the usefulness of
intervention from "outside" for local sustainable development, and, indeed,
the usefulness of political and economic integration of rural communities
more generally. In the Indian context, these neo-populists often refer to
Mahatma Gandhi's concept of self-sufficient village republics. According
to them, because of India's cultural and biological diversity, ecological
regeneration can only be achieved through decentralized decision-making
and village autarky (Agarwal & Narain 1993).

However, conflicting interests, heterogeneity and inequality within
communities may hinder participatory initiatives for sustainable development. Furthermore, as political ecology has pointed out, the wider political
and socioeconomic context finally determines whether the indigenous
potential for development can be made use of. Yet, community-based
development has drawn our attention to local conditions and practices, and
indicated that "imported" technologies and knowledge systems can be
possible causes of environmental degradation.

2.6   Theoretical Complementarities and Differences

The discussed approaches point to important causes of environmental
degradation (apart from causes such as population growth and unsound

consumption-oriented lifestyles). Neoclassical environmental economics has concentrated on misallocation of resources, price distortion, and on reformist policy instruments that are practicable in the short term. By contrast, political ecology and community-based development addressed "big" causes of environmental degradation such as unequal access to resources or inappropriate technologies; and they suggested that more fundamental change is necessary to achieve sustainable development.

The theories I have discussed are complementary to a certain degree; but they also indicate differences with respect to the relation between crop markets and environmental problems. Neoclassical environmental economics suggests that markets may have positive or negative effects on the environment, depending on whether or not prices reflect environmental costs. By contrast, political ecology seems to imply that capitalist relations of production and markets compel poor peasants in the Third World to overuse natural resources and thereby degrade the environment. Finally, community-based development implicitly refers to markets as mediums of outside intervention that tend to displace indigenously developed, sustainable agricultural systems.

One reason for the contrary assessment of the impact of crop markets is that the various approaches to sustainable development refer to different underlying development theories and concepts of market exchange. Furthermore, the impact of crop markets on the environment seems mostly indirect; that is, through their influence on agricultural practice and socioeconomic development. In order to better understand the relationships between agricultural markets and sustainable development, the next chapter thus introduces broader concepts of markets and agricultural change.

3 Markets and Agricultural Change

Agricultural markets, state interventions, agrarian relations, infrastructure and technology and their relationship to development have been widely discussed in development theories over the past five decades. Neoclassical, neo-Marxist and new institutional approaches have suggested particular concepts of market exchange and models of agricultural development or stagnation. These development theories have concentrated on different aspects of agricultural markets, and have generated various hypotheses about their influence on socioeconomic development that, if reinterpreted in the concepts of social practice and real markets, are relevant for this study. In order to facilitate this theoretical discussion, this chapter first explains the characteristics of agricultural practice and of agricultural markets in developing countries.

3.1 Agricultural Practice

As cultivators are principal agents of agricultural change, their practices and decision-making are paid special attention. Various theories about agricultural decision-making in developing countries have characterized cultivators in different ways. Whether peasants are "rational" or not and to what motives or rationales the cultivators' "rationality" relates have been subject to debate. Schultz (1964) attacked the long prevailing, simplistic notion of peasants' irrationality most vigorously. Within the limitations of "traditional" agriculture, Schultz regarded farmers as rational, *profit-maximizing* entrepreneurs. In more recent theories, the farmers' rationality has not been equated with profit maximization. Rational-choice theory, for example, relates rationality to a means-ends relationship in which the end is expected "utility," which can be anything from profit to leisure maximization.

Lipton (1968), furthermore, argued that for peasants in the Third World, who produce under conditions of uncertainty (e.g., natural hazards, limited information, price fluctuations, social insecurity and political instability), *risk minimization* may be the "optimal," "rational" strategy. Profit maximization, by contrast, involves too high risks and can lead to an extremely severe outcome (i.e., starvation). Farmers in developing countries, therefore, may seek survival algorithms, not maximizing ones. Furthermore, different farmers learn and stick to different algorithms, which also explain wide inter-farm differences in the use of similar resources.

However, concepts of peasants' rationality presume utilitarianism, narrow self-interest and exclude the possibility that altruism, cooperation or protecting the environment can also be motives or ends in themselves (as opposed to mere means for utility maximization). They also imply that means and end would always be clearly distinguishable. To assign economic rationality to peasants and to their every single action, furthermore, seems as inappropriate as to use simplistic, generalizing notions of peasants' ignorance. Peasants, like other agents, are sometimes, and to a degree, "rational" (J. Adams 1986: 274). Instead of making assumptions regarding peasants' rationality, therefore, it is more appropriate to regard peasants as knowledgeable and capable without presuming that they always make rational decisions; peasants often conduct conventional, routinized practices and cultivation habits.

The importance of individual motives and attitudes should not be over-estimated: individuals are influenced by social norms, cultural values and larger frames of meaning. Nevertheless, the motives of individuals may still partly explain variation from cultivator to cultivator regarding cropping patterns and cultivation methods. On the other hand, cultivators seldom have clear-cut motives. Nor does each cultivator possess a particular mix of motives that applies to all his or her agricultural practices. For example, the various practices of a single cultivator (e.g., the cultivation of different crops) may be based on different mixes of profit-, security-, leisure-, and status-maximizing, altruism, and aesthetic or environmental motives. Instead of assuming farmers to be profit or security maximizers, it might therefore be more useful to study for what kind of agricultural practices farmers tend to have profit- or security-oriented motives.

Whatever the motives of an individual and to whatever degree they are influenced by sociocultural factors, motives alone do not determine agricultural practices. Peasants may well have profit-maximizing attitudes, for example, but they may not be *able* to cultivate a remunerative crop because they lack suitable land and water resources, investment capital, or market access.

> [F]armers' failure to innovate or to increase production for the market often reflects their inability rather than their unwillingness to do so.... In brief,

farmers' readiness to take advantage of new income earning opportunities often depends more on their assets than on their attitudes. (Berry 1980: 329)

For the study of the relations between markets and sustainable development, it is, therefore, necessary to acknowledge social differentiation of cultivators according to class, caste, ethnicity and gender.

3.2   Agricultural Markets

As markets can be enabling or restricting conditions for farmers' social practices, they have also an effect on agricultural development. Most development theories, however, have used an abstract concept of the market that neglects qualitative aspects and social embeddedness of markets. The abstract market concept is in line with mainstream economists, who define the market as "an abstract pricemaking mechanism that is central to the allocation of resources in an economy" according to supply and demand (Swedberg 1994: 255). Although also monopoly and oligopoly are considered as possible transactional conditions, the market is normally understood as autonomous and flexible mechanism of exchange based on choice and perfect competition. "The institutions of the market are assumed to be independent, atomistic and impersonal.... It is [Adam] Smith's invisible hand" (B. Harriss 1991: 3).

The mainstream economists' formal, abstract concepts have been contested by *substantive economics* (Polanyi 1968), which calls for economic analysis that is not derived from logic but from empirical fact. According to Mackintosh (1990: 47), actual processes of buying and selling have been investigated mainly by anthropologists and geographers. These empirical studies have revealed a very wide range of different markets, which have little in common with the complete, perfect market of formal economics, but rather are unique in space and time. Polanyi himself also suggested that not only the market, but also redistribution and reciprocity have to be considered as operating exchange mechanisms.

Particularly in developing countries, markets may remain incomplete (i.e., not fully developed) and they operate under conditions of limited information, imperfect competition, and asymmetric power relations. Furthermore, according to Harriss-White (1996b: 29, 39), markets are embedded in:
-   social structures (e.g., relations of class, caste, ethnicity, kin, gender; property-right regimes; distribution of ownership);

- the state (as a law-making regulator or as a public-enterprise-managing competitor);
- technical-material properties (e.g., soils and climate, infrastructure, physical nature of the marketed commodity).

These *real markets* are not comprehensible as impersonal mechanisms of free exchange based perfect competition. Rather, in this study, markets are defined as institutionalized interactions between sellers and buyers that include exchange of particular commodities or services at a mutually agreed (but not necessarily mutually beneficial) exchange rate under conditions of rivalry or a degree of competition. This definition of markets includes both a vertical component, which describes interactions between buyer and seller, and a horizontal component, which refers to structural properties and economic conditions of market exchange. Market interactions often involve formal and informal regulations, contractual arrangements, and either personal relations between buyer and seller, or more remote – often unequal – social relations between them. Market interactions usually are repeated and often build up marketing chains and networks for particular commodities that stretch over time and space. The horizontal component of markets refers to structural properties and general conditions such as numbers of buyers and sellers, degree of competition, seasonal and spatial demand-and-supply structures and prices.

Agricultural markets are diverse and highly complex. They include crop markets and agricultural-input markets for land, labor and capital. Agricultural marketing is also more than just "trading" as it involves productive activities: assembly, post-harvest treatment, storage, processing and transport. In India, furthermore, the often very high concentration of economic assets stands for power relations and behavior such as stock holding, bargaining advantage, pressing decisions, controlling information, constraining the choices of others, linking and/or segmenting markets in the contractual encounter (Harriss-White 1996b: 30-31). Moreover, local, rural markets may be financed, organized and spatially structured by urban wholesalers. Petty trade in weekly food markets can become a dependent sub-system of a dominant encompassing market system (Bohle 1990). Besides, even in the same locality, local markets may coexist with global markets.

The neoclassical, new institutional and neo-Marxist theories discussed in the following three sections emphasize different aspects of agricultural markets that influence agricultural practice and development.

3.3   Price Signals

The power of price signals has been a main pillar of neoclassical theories of agricultural development. Relative prices are believed to determine agricultural practices, and to influence technological and institutional change. The use of comparative advantage through "free" trade would lead to maximal efficiency and social benefit. Despite its unrealistic assumptions, neoclassical economics has also had significant impact on development practice.

3.3.1   Relative Prices and Comparative Advantage

The impact of relative prices on agricultural practice (cropping pattern, production intensity and cultivation method) has been addressed by the theory of *farm production*. The farm-production model has concentrated on farmers as individual decision-makers who are concerned with such questions as what combination of crops to grow and how much (purchased) inputs (land, labor, capital, fertilizer, etc.) to devote in what combination to the cultivation of each crop. The theory assumes that farmers are profit maximizers who have complete information. Among many other assumptions of neoclassical economics, particular assumptions related to the farm-production model include perfect competition in markets for farm inputs and outputs; unlimited working capital for the purchase of inputs; substitutability of inputs and of outputs (Ellis 1988: 17). Derived from these assumptions, neoclassical theory predicts that farmers would:
- produce a crop combination from a given set of farm resources so that the marginal returns for each crop are equal (product-product relationship; optimal enterprise choice);
- use an amount of inputs for which the extra return is equal to the extra cost for the marginal unit of input used (input-output relationship; optimal production function);
- combine the inputs in such a way that the marginal product per unit spent is the same for each input (input-input relationship; optimal method of production) (Ellis 1988: 16-17).

In brief, cultivators are assumed to allocate their inputs efficiently, produce the most profitable mix of crops, intensify production optimally according to economic calculations, and apply the least-cost method of production. In this neoclassical model, farm production is therefore a function of *relative prices* among alternative agricultural products and different agricultural inputs. Given the farm's resources, crop markets (i.e., relative prices for agricultural products) would influence what kind of crops a cultivator will grow and whether he or she will intensify production. Furthermore, relative prices (and therefore agricultural production) may be affected by a rise in population that leads to greater demand, and also by technological change.

As peasants (and other agents) are not sheer profit maximizers, relative prices may not have such a direct impact on agricultural practice as neoclassical theory suggested. Furthermore, it is unrealistic and misleading to reduce input and output markets to prices, and to assume that agricultural markets in developing countries are perfect, autonomous and power neutral. The influence of relative prices on agricultural practice, therefore, remains an empirical question for which agricultural markets have to be studied as part of agrarian relations and in their social and political context.

On the societal level, neoclassical economics emphasized the significance of *comparative advantage* for agricultural growth. The principle of comparative advantage refers to the spatial distribution of inputs for production. Region A, for example, may have a comparative advantage (i.e., comparatively low production costs) over region B for growing paddy, because of climate, soils or seasonal labor availability; meanwhile region B may have a comparative advantage over region A in the production of coconut. Neoclassical theory suggests that for a society (or the world economy) to become economically efficient, cultivators, regions and countries should specialize production according to their respective comparative advantage, and then exchange the products in the market. For example, region A would trade paddy against coconut with region B. In order to make full use of comparative advantage, exchange of agricultural commodities should not be constrained in any way. *Free trade* is believed to result in advantages for all trading parties and maximize economic development of society as a whole.

However, by demanding free trade in every case, neoclassical theory neglects transaction costs (see Section 3.5). Also, trade between A and B (whether regions or individuals) may leave C worse off. Just as there is comparative advantage, there is comparative disadvantage (B. Harriss 1991: 12). For development in Third-World countries, moreover, long-term institutional change may be more fundamental than the "right" short-term allocation of resources. Although neoclassical economics has been widely criticized for its many unrealistic assumptions and for its neglect of core issues of development, it has had significant impact on development practice – particularly regarding the modernization strategies initiated in the

1950s and the neoliberal structural-adjustment programs popular since the 1980s.

3.3.2 Modernization and Market Integration

Regarding "planned" agricultural development, T.W. Schultz's publication *Transforming Traditional Agriculture* (1964), which adheres to neoclassical economics, has been a major intellectual source. Schultz vigorously attacked the formerly popular thesis of peasants' ignorance and irrationality. On the basis of two monographs (one of which was about Indian farmers), he concluded boldly that "there are comparatively few significant inefficiencies in the allocation of the factors of production in traditional agriculture" (Schultz 1964: 37-38). Cultivators and entire "traditional" farming communities in developing countries seem to be *efficient but poor*. This allocative efficiency may have been achieved over generations in allegedly constant, "traditional" environments. However, Schultz made a distinction between allocative and technological efficiency. He believed traditional agriculture to be technologically "backward." In order to transform the allegedly backward, traditional agriculture, he advocated transfer of foreign "scientific" know-how and agricultural research.

Schultz's use of empirical evidence has been criticized as highly selective and representing rather untypical situations. Schultz's interpretation of the two chosen monographs was also biased unfairly to favor his thesis of allocative efficiency in farming communities (J. Adams 1986: 275-276). Furthermore, indigenous-knowledge studies have challenged the belief that "scientific" research stations alone are appropriate to develop agricultural technology for developing countries. Despite this, the idea of transforming and modernizing traditional agriculture was widely implemented in the 1960s with the so-called *Green Revolution*, which advocated complete technological packages of purchased inputs (new seed varieties, agro-chemicals and irrigation). The Green Revolution focused on monocultures, involved large-scale irrigation of plains, and neglected dryland farming (Chambers *et al.* 1989).

The second pillar of agricultural modernization was to integrate "traditional" communities into the national market economy so that the efficient farmers would have the opportunity to make use of comparative advantage. In the 1970s, the commercialization of agriculture became an important element of the World Bank's *Rural Development* strategy. In a "modernized" agriculture (characterized by "modern" farm inputs and market integration), farmers were believed to become commercial calculators who would consider input and output prices and would react to price signals according to the neoclassical model of farm production. The state was ascribed the leading role for the development and extension of agricultural technology, construction of irrigation facilities, building up of marketing facilities and also interference in markets such as subsidizing rural credit, supporting marketing cooperatives and fixing commodity prices.

The record of agricultural modernization and commercialization was mixed. Populists and neo-Marxists have shown that only rich farmers were able to benefit from the Green Revolution, and that commercialization enhanced the uneven integration into markets (see Section 3.4). Neoliberals, on the other hand, have focused on state failures and the negative effects of state development intervention (Lal 1983). They have been outspokenly anti-interventionist, have trusted in free-market forces and, thus, advocated *structural adjustment*. In particular, the state was advised to:

- focus on a few key areas (e.g., infrastructure, protection of private property rights);
- turn away from import-substitution strategies;
- remove related market restrictions (which seem to be biased against agriculture and in favor of industry) and open up the economy;
- remove price controls *(getting the prices right)* (Balassa 1982; Lal 1990).

The neoliberal advocates of structural adjustment assumed that, without the interventionist state, markets work under relatively "perfect" competition. Taking this assumption of the neoclassical model for reality, they expected that structural adjustment would result in comparative advantage for the farmers in developing countries and, therefore, lead to agricultural growth. In some ways, the neoliberal orthodoxy can be interpreted as a return to the colonial model of comparative advantage and export-oriented commodity production, or as a *back-to-the-future* strategy (Peet & Watts 1993: 236).

In addition, neoliberal theory predicted that because of subsidy cuts on energy and agro-chemicals, structural adjustment, although it has been designed for other reasons, would also contribute to environmental protection. A comprehensive study about the impact of structural adjustment on the environment, however, concludes that this impact is likely to be random and mixed, positive and negative (Reed 1992).

3.3.3 Induced Technological and Institutional Change

Neoclassical theory is also the basis of mainstream theories of agricultural development such as the theory of *induced innovation* (Hayami & Ruttan 1985). This neoclassical institutional theory has incorporated technological and institutional change into a comprehensive economic model. According to the theory of induced innovation, technological change (like institutional change) is influenced by the resource endowment (land and labor) of a particular society and a growth of demand that results in changes of relative prices:

> [F]armers are induced, by shifts in relative prices, to search for technical alternatives that save the increasingly scarce factors of production. They press the public research institutions to develop new technology and also demand that agricultural supply firms supply modern technical inputs that substitute for the more scarce factors. (Hayami & Ruttan 1985: 88)

For agricultural development, the dominant technological substitute in the case of land scarcity may be bio-chemical technology (e.g., fertilizers and pesticides, high-yielding crop varieties, irrigation); mechanization, on the other hand, may substitute for scarce labor. Generally, technical change becomes cheaper (and thus more likely to happen) with the general progress of science. However, technical change may only be guided along an "efficient" path, if factor and product prices really reflect the demand and supply of factors and product, and if there is effective interaction between farmers who are politically organized, on the one hand, and scientists and administrators who are adequately rewarded for their contribution to the solution of problems of the society, on the other hand (Hayami & Ruttan 1985: 73-84).

Similarly, institutional innovation is seen as induced by changes in technology, resource endowment and product demand. In some cases, the "demand" for institutional innovation can be met by the development of new forms of property rights or more efficient market institutions. However, such institutional development requires substantial political resources (i.e., "costs"). For a society, institutional change is profitable only if "benefits" are higher than "costs." Otherwise, the continuation of "traditional" institutions would be more appropriate. Particularly for the supply of public goods to a large, unidentifiable clientele, the market as a "modern" institution tends to fail (Hayami & Ruttan 1985: 103-107).

However, actual institutional change is not necessarily congruent with increased efficiency. Powerful interest groups and organizations are able to impede economically desirable institutional change when they fear to lose their privileges. Likewise, they are able to push socially undesirable change if it is specifically to their benefit (Hayami & Ruttan 1985: 107-108; North 1990: 7-8). Consequently, the "supply" of institutional innovation can be limited ("made costly") by the power structure among vested interest groups. Furthermore, the supply of institutional innovation is dependent on the *cultural endowment* (i.e., cultural tradition, religion, ideology). Moral obligations and traditional patterns of cooperation, for example, may represent an important foundation on which modern forms of cooperation such as joint farming or cooperative marketing can be erected. However, such cultural "resources" are not believed to be available in South Asia, where caste structures inhibit cooperation and encourage specialization. (Hayami & Ruttan 1985: 107-110).

The theory of induced innovation has been criticized for using economic jargon and emphasizing economic factors as the main reasons for institutional change. According to Grabowski (1988), economic factors explain only changes in "secondary" institutions (e.g., property rights, tenancy rules, marketing arrangements) but not so much the changes of fundamental institutions (i.e., systems of cultural values and beliefs). Indigenous-knowledge studies have also challenged the notion that public research institutions are the main innovators of agricultural technology (see Section 2.5); and neo-Marxists have shown that, because of unequal agrarian relations and the possibility of investing in non-agricultural activities, the introduction of new agricultural technology does not always stimulate agricultural growth (see Section 3.4).

3.4 Uneven Market Exchange

The effects of uneven exchange in markets have been discussed most rigorously by neo-Marxists. Unlike neoclassical economics, neo-Marxist theory focused neither on the influence of relative prices nor on problems of "optimal" resource allocation, but rather on the actual effects that "imperfect" markets have on development and economic well-being of different social classes. Neo-marxists considered the whole economic, social and political context (mode of production) in which markets work. *Mode of production* describes the dominant, total system of social and economic organization of a society (e.g., feudal mode of production, capitalist mode of production). It includes *social relations of production* (i.e., the social organization of access to and of control over *means of production*), *forces of production* (the material basis including technology) and the *superstructure* (i.e., legal and cultural norms). As a consequence of its distinct research focus, methodology and ideology, neo-Marxist theory indicates that under the existing exploitative agrarian relations of production in developing countries and in the world economy, commercialization only perpetuates exploitation and inequality.

3.4.1 Agrarian Relations of Production

Unlike the Latin American debate about dependency and exploitative exchange relations in a world economy (e.g., Frank 1969), Indian neo-Marxists in the 1970s and early 1980s engaged in a debate about modes of production and their consequences for agricultural development. This debate is basically about whether Indian agriculture is characterized by a capitalist mode of production, a particular pre-capitalist mode of production or an articulation of several modes or forms of production (Blomström & Hettne 1984: 128). By contrast, *orthodox Marxism* predicted a complete transition from the feudal mode of production to the capitalist mode of production, which would be pre-condition for socialist revolution. This transition would imply that the agrarian structure differentiated the peasantry into capitalist farmers and rural wage laborers (proletarians).

The *neo-Marxist* mode-of-production debate provided various explanations why, in reality, this differentiation has not happened completely, or why it has been muted. Non-Marxists, too, have sought to explain the persistence of the middle peasantry. They have based their arguments on such things as the distinct social norms of peasant communities *(moral economy,* Scott 1976) and the peasants' capacity for self-exploitation in face of market pressure (Chayanov 1966 [1925]). More recently, Mann (1990) has argued that capitalism failed to dominate agricultural production as much as industrial production because of such natural peculiarities as long production time and seasonally fluctuations in labor requirements. Together, these make farming risky, unprofitable for capital investment, and unsuitable for the use of wage labor.

A first group of modes-of-production theorists have argued that the world economic system may not be marked by capitalism alone (as dependency theory proposed) but by elements belonging to different modes of production. Indeed, they maintain that capitalism, rather than eliminating pre-capitalist relations in agriculture according to its own "needs," may actually reinforce or even restore them (Laclau 1971). In India's agrarian structure, according to Bhaduri (1973, 1986), *semi-feudal relations* may persist in articulation with capitalist structures. In this theory, landlords occupy a central position. The landlords may exploit their tenants and sharecroppers via land rent and usury. Usury may consist of consumption credit extended in the off-season and of lending seeds and other agricultural inputs at the beginning of the agricultural season. In order to ensure the persistence of semi-feudal production relations and surplus appropriation, landlords may attempt to keep their tenants indebted and thus may have no interest in introducing organizational or technological improvements that would raise the tenants' income (Bhaduri 1986: 269).

The second group of the modes-of-production school distinguish between European feudalism and the *colonial mode of production:* the introduction of "feudal-like" tenancies in response to long-distance trade in the colonies of the Third World (Banaji 1972; Alavi 1975; both cited in Booth 1985: 769). In the colonial, as well as post- or neo-colonial, mode of production, capitalist surplus appropriation of wage labor may coexist with various other relations of exploitation. Peasants, for example, may formally possess the means of production, but in reality, their labor may be subsumed under capital, particularly merchant capital. According to this theory, not landlords but merchants take the leading role as exploiters. Merchants can function as moneylenders and extend consumption and production loans to the peasants. In this way, the merchants-cum-money-lenders may establish control over agricultural production in a monetized and "commoditized" rural economy. Consequently, the merchants may also be able to forestall technical development. Furthermore, the combined merchant/usurious capital connects petty commodity production to the capitalist world market (Banaji 1977, cited in Athreya *et al.* 1990: 237-238). However, empirical studies show that the mercantile sector does not *necessarily* function to promote unproductive accumulation, preserve pre-capitalist production relations and forestall technical development; it may also promote productive accumulation, capitalist relations and technical change (B. Harriss 1989).

The third position within the modes-of-production school proclaims that pre-capitalist *forms* of production may persist within the capitalist *mode* of production in the Third World. In each historical period, various forms of production coexist in different relations to each other. Pre-capitalist forms would follow the logic of subsistence, as opposed to capitalist appropriation of surplus-value, capital accumulation and invest-ment. The penetration of the capitalist mode of production and the moneti-zation of the economy, however, initiate a process of *commoditization* that disrupts the autonomy of the reproduction process. Households become dependent upon capitalists for their reproduction. They produce not only for direct household consumption (use-values) but also for sale (commodities, exchange-values).

However, the simple commodity producers, because they are both production and consumption units, still follow the subsistence logic. They are primarily concerned with producing their own livelihood – and with reproducing their means of livelihood. Therefore, they continue to produce in circumstances in which capitalist firms would go out of business because of inadequate returns on investment. Peasant households and other household producers may even be forced to produce more when prices fall in order to meet the needs for simple reproduction. Similarly, in order to secure a plot of land for subsistence, they may pay higher than average rents. This mechanism has been labelled *simple reproduction squeeze* (Bernstein 1982); it represents a reformulation and an extension of the concept of the peasants' capacity for *self-exploitation* (Chayanov 1966 [1925]), which did only refer to peasants and tended to imply the exploitation of women, children and elderly people within a household rather than "self-exploitation." The concept of the simple reproduction squeeze stands in diametrical opposition to the less realistic neoclassical model (i.e., the production function), which supposed that when prices fall and extra returns thus fall below the costs of marginal inputs, farmers would reduce inputs (e.g., labor). The simple reproduction squeeze enables the capitalist sector to extract labor and commodities below their value from pre-capitalist sectors. In other words, the peasant sector (and the informal urban sector) subsidize the formal sector and are integral parts of the capitalist mode of production (see Evers 1987).

The important implication of this perspective is that capitalists (merchants or industrialists) may be able to regulate the conditions of peasant production and exchange without undertaking its direct organization (Bernstein 1982: 165). Some Indian scholars expect that as a consequence of India's current economic liberalization, the influence of industrialists (particularly transnational companies) on agricultural development is likely to rise. Monetization and commercialization of agriculture may now be followed by the industrialization of agricultural production and its subsumption under the international agro-industrial chain of production. This process may also include the emergence of large-scale corporate farming and contract-farming, which, according to these scholars, would put pressure on already marginalized sections (Krishnaswamy 1994).

3.4.2 Unequal Power and Forced Commerce

Although neo-Marxist theory focused on relations of production and surplus appropriation labor, some scholars have also discussed market exchange and relations of circulation. Bharadwaj (1985) studied different conditions of market involvement and the actual effects that market exchange has on different social classes of peasants. Bhaduri (1986) paid attention to the *class efficiency of markets*, the power of particular classes to appropriate surplus through the market.

Neo-Marxist theories reject the idea of "perfect" markets in which all types of peasants would be free to engage or withdraw. In reality, the character of markets may reflect, and be determined significantly by, local patterns of power. Moreover, commercialization and exchange can re-inforce these patterns of dominance and power. Thus, the involvement in the market of different types of cultivators shows principal differences and depends largely on their initial resource position and their resulting bargaining strength: Big landlords, moneylenders and merchants are able to dominate the markets for land, leases, credit and agricultural products.

Beyond monopolistic control over prices, dominant classes may thus also be able to determine the terms and conditions of contracts and to shape the character of exchange relations in their favor. Moreover, single persons often combine the functions of landlord, moneylender and merchant and, thereby, "interlock" various input and output markets contractually. These *interlocked markets* increase the exploitative power of the dominant classes. While there might be conventions and economic limits of exploitation in any single market, interlocked markets create a situation in which the dominant party controls the entire livelihood of the weaker party, especially when the latter is indebted. Economically very weak people – landless agricultural laborers, very small owner-cultivators or tenants – may not be able to retreat from market operations whenever some of the interlocked markets work to their disadvantage *(compulsive involvement in markets)*. They can be forced to produce cash crops, to hire out their labor, to lease in land at high rents, and to take consumption and production loans. Small and middle peasants, on the other hand, may be able to protect themselves partially from the vagaries of markets and the dependence on landlords-cum-merchants-cum-moneylenders (Bharadwaj 1982: 271-274; Bharadwaj 1985).

According to neo-Marxist theory, therefore, crop markets are not primarily characterized by relative prices but by exploitative relations that must be analyzed together with the whole set of agrarian relations, which reach into land, labor and credit relations. Bhaduri (1986) understands market exchange as a mechanism of surplus extraction, rather than a means to make use of comparative advantage that would be beneficial for all parties, as the neoclassical model suggests:

> [T]he function of exchange is not to "clear" the market in some cases, but simply to give advantage to one party *at the cost* of another. Exchange relations, under these circumstances, are better understood as a mechanism for surplus extraction. Consequently, the "market mechanism" is also better understood not in terms of its allocative efficiency, but as the mechanism for extraction of surplus by one class from another. In the context of backward agriculture, I coined the term *"forced commerce"* to emphasize precisely this possibility.... Under such forced commerce, exchange relations may be "efficient" from the point of view of the surplus-extracting class, quite irrespective of whether or not they satisfy the neoclassical postulates of allocative efficiency. (Bhaduri 1986: 268)

Neo-Marxists may overstress the exploitative character of markets while neoclassical economics overemphasize mutual benefits through exchange. However, the concepts of forced commerce and class efficiency of markets are useful sensitizing devices for this study as they point to the necessity of taking into account agrarian class relations and the wider economic, social and political context of agricultural production and exchange.

3.4.3 *Commercialization and Agricultural Development*

Forced commercialization is likely to accentuate the existing power inequality and differentiation among people who enter the market on unequal terms. To the extent that markets are interlocked and market involvement diverges qualitatively between different types of cultivators, market integration induces disparate, iniquitous results for different types of cultivators. The experience of India's early commercialization in the 19th century indicates that commercialization of the agricultural-product market can reinforce exploitative, pre-capitalist land, labor and credit relations (Bharadwaj 1985: 334). Usually, not the poor but the relatively well-to-do farmers are able to get access to new global or national product markets as well as to state-subsidized credit and technology. Therefore, neo-Marxists have criticized the attempts of the modernization and structural-adjustment programs to commercialize India's agriculture and to integrate peasants into national and global economies without tackling existing agrarian relations.

Doing so would only enhance the opportunities for accumulation by the already rich peasants, while the many poor peasants might become even more vulnerable, impoverished and marginalized (Bernstein 1990: 75-76).

Because the dominant classes tend to reinvest the surplus extracted from agriculture in "unproductive" activities such as usury, speculation and trade, the persistence of uneven market exchange not only causes growing disparities but ultimately may also hinder productive investment in agriculture and therefore agricultural growth (Bharadwaj 1985: 338-339). Because investment in non-agricultural activities is more profitable, the introduction of new agricultural technology may stimulate agricultural growth only initially and for a limited period of time. Furthermore, even the redistribution of land under land reforms may not be able to induce agricultural growth. Although small farmers tend to be the most efficient (mainly because of their capacity of "self-exploitation"), the redistribution of land may not be sustainable because new small landowners lack production capital, become indebted, mortgage their properties, and eventually lose their land again (Bharadwaj 1985: 345).

The pessimistic neo-Marxist scenario of growing inequality and agricultural stagnation has been criticized on theoretical and empirical grounds. From a meta-theoretical perspective, Marxism tends to be inappropriately deterministic, functionalist and teleological. Consequently, neo-Marxist development theory, instead of showing why the world is the way it is and how it may be changed, tries to demonstrate that "structures and processes that we find in the less developed world are not only explicable but *necessary* under capitalism" (Booth 1985: 776; emphasis not in original). An exception to Marxist determinism is Bernstein (1982), who mentions the possibility of different development paths and points to the dangers of facile generalization. Generally, neo-Marxists overstress constraints imposed on cultivators while neoclassical economics over-emphasize the autonomy of cultivators. Dependency theory excluded *a priori* that global and national markets may also form opportunities for cultivators.

More recent Marxist-influenced development theories have moved away from structural determinism; *post-impasse theories* (Schuurman 1993; Booth 1994) have overcome the meta-theoretical impasse of neo-Marxist development theory identified by Booth (1985). They give due attention to the *room for manoeuvre* of agents and states within wider, diverse structures, and consider the possibility that the circle of underdevelopment can be broken. On a global scale, furthermore, the relation between countries is increasingly perceived as one of interdependence rather than of unilateral dependence.

In addition, recent empirical studies in India have overcome both economistic reductionism and the focus on *local* power structures inherent in the Indian mode-of-production debate. By contrast, these studies consider the influence of politics, culture and ideology on agrarian change (e.g., J. Harriss 1994; Athreya *et al.* 1990). In Tamil Nadu, for example, state interventions contributed crucially to agrarian change and growth without further marginalizing the poor sections of society. Consequently, the neo-Marxist view of the state as a mere protector of existing relations of production and local agrarian structures needs to be replaced with a more complex view (Athreya *et al.* 1990: 313-314). In a democracy such as India, demands of peasants can influence policies and state interventions too, especially when the peasants are organized. Particularly in the past decade, India has seen the rise of powerful farmers' movements that ask the state for higher crop prices, subsidies for agricultural inputs, etc. (J. Harriss 1994: 174). Furthermore, Indian environmental movements have emerged that aim at protecting people's livelihood. These movements demand more equal access to resources as well as a halt to large-scale, environmentally and socially unsound development projects.

3.5   Costliness of Market Exchange

Ensminger (1992) has argued that *new institutional economics* provides an alternative approach to both neoclassical and Marxist explanations of agrarian change and development. Unlike neoclassical economists who believe that commercialization *automatically* leads to economic growth and to benefits for all, or neo-Marxists who suppose that trade under the prevailing imperfect, exploitative markets *inevitably* leads to increased inequality and underdevelopment, new institutional economists have argued that both are possible but neither is necessary or inevitable. These theoretical differences arose because the new institutional theory has considered the costliness of exchange (as opposed to neoclassical economics) and the possibility that institutions and power relations can be altered (as opposed to neo-Marxist theory).

3.5.1 Transaction Costs

New institutional economists have criticized economic theory for failing to take the costliness of exchange into account:

> For 200 years the gains from trade made possible by increasing specialization and division of labor have been the cornerstone of economic theory. Specialization could be realized by increasing the size of markets, and as the world's economy grew and division of labor became ever more specific, the number of exchanges involved in the performance of economies expanded. But the long line of economists who built this approach into an elegant body of economic theory did so without regard to the costliness of this exchange process. (North 1990: 27)

While transportation costs have partially been considered in economic theory, transaction costs, which also contribute to the costliness of exchange, have been neglected. *Transaction costs* include costs related to the search for information, calculation of risks, negotiations for transactions and the development of mechanisms of monitoring and enforcing contracts. These costs evolve because, in transfers, it is necessary to define, measure, protect and enforce property rights to goods: the right to use, the right to derive income from the use of, the right to exclude and the right to exchange (North 1990: 28). As market exchange involves high transaction costs under conditions of imperfect information and pervasive risks, new institutional economists have argued that non-market institutions are often more efficient than markets.

Within communities, transaction costs of exchange tend to be low because of:
- predominance of traditional forms of exchange based on indigenous institutions such as reciprocity and redistribution;
- personalized forms of market exchange in which individuals are known (if not related) to one another and engage in repeat dealings.

However, potential gains from small-scale exchange within communities are highly limited because the trading partners are likely to face the same environmental conditions and often produce the same products. In order to capture higher comparative advantage, therefore, broader trade and specialization are necessary, yet this also implies an increase in transaction costs. When the additional transaction costs of the exchange process outweigh the gains from exchange, societies do not benefit from trade and specialization (Ensminger 1992: 25).

Exchange costs are not fixed. They may be reduced through infrastructural development and institutional innovation. New institutional theory has stressed the crucial role of the state in providing marketing *infrastructure* (e.g., roads, telecommunication) and in establishing particular *institutions* to reduce transaction costs. According to Ensminger (1992: 27), the state plays an ambiguous role with respect to such institutions. On the one hand, the state reduces transaction costs through:

- defining property rights;
- providing regulations regarding weights and measurements;
- acting as a third party of legal enforcement of contracts and property rights.

On the other hand, the state can also increase transaction costs through:

- monopolization of agricultural marketing;
- excessive regulation and licensing;
- corruption in courts and police.

Furthermore, indigenous institutions, personalized transactions and cultural values continue to be important for reducing transaction costs – even in a society with relatively developed and specialized modern institutions.

In developing countries, exchange costs tend to be high. Market institutions designed to facilitate the free flow of information on supply and demand and to adjudicate disputes over property rights, trading contracts, loan repayments, etc. are often unreliable or nonexistent. As a consequence, particular marketing arrangements (what new institutionalists have called *non-market institutions)* may be employed to reduce the transaction costs and uncertainty for agricultural marketing. In order to facilitate exchange in situations of undeveloped "modern" institutions, vertical integration (i.e., the involvement of one organization or family in different stages of production and distribution of a particular commodity) or the organization of marketing through a trading diaspora may be applied (Ensminger 1992: 26-27).

As a consequence of such marketing arrangements, small traders might find it difficult to be competitive in any segment of the marketing system (Ensminger 1992: 27). Empirical evidence, however, suggests that petty traders can coexist with dominant traders of a trading diaspora. This phenomenon can be explained with the *moral economy* (Scott 1976) within villages that gives rise to the *trader's dilemma* (Evers & Schrader 1994). Because transactions within the community are governed by a moral economy of fairness that inhibits traders from exploiting fellow villagers, traders from peasant societies are faced with a difficult choice between either losing money or losing social esteem. There are two solutions to this problem. First, trade becomes the province of a distinct ethnic group that is set apart from the village norms and can therefore accumulate capital necessary for large-scale and long-distance trade. Second, trade is limited to the cash-and-carry activities of petty traders, often women, whose profits are self-evidently so low as to deflect any pressures for sharing or redistribution (Evers & Schrader 1994).

Probably more common than vertical integration is the contractual interlocking of agricultural markets, particularly pre-harvest purchases (i.e., the interlocking of a pre-harvest money market with a post-harvest commodity market). New institutional economists have so far not focused on such arrangements in crop markets but rather on agrarian land and labor relations under circumstances of imperfect information (B. Harriss 1991: 9). Sharecropping, for example, has been interpreted as the most efficient institutional form because, in environments of high risks, asymmetric information and moral hazard, sharecropping tends to involve lower monitoring and enforcement costs than the employment of wage labor (Stiglitz 1986; Popkin 1988). Generally, new institutional economists emphasize mutual interest and mutual safeguarding commitments. They interpret arrangements such as sharecropping or pre-harvest purchases as reducing uncertainty for all the involved parties (landlord, cultivator and trader).

However, sharecropping, the interlocking of factor and product markets and particular agricultural-marketing arrangements, like all institutions, should not be seen as *automatic* outcomes ruled by a "law of efficiency." Particular institutions are also shaped by interaction under conditions of asymmetric *power* (Bardhan 1989) and by unintended consequences of interaction. Preexisting power relations often determine whether higher agricultural income, attained from institutional (or technological) innovation, increased specialization and trade, is distributed to all groups of a society or whether it leaves some groups the same or even worse off (Ensminger 1992: 28).

Its attention to transaction costs and various marketing arrangements makes the new institutional theory a useful sensitizing tool for this study. This will only prove effective, however, if sufficient attention is paid to power relations and the concept of transaction costs is interpreted in a non-deterministic way, and if outdated assumptions about the "rationality" of individual behavior and the "efficiency" of institutional development are replaced with the concept of social practice.

3.5.2   *Indigenous Institutions and Community-Based Organizations*

For development to occur, not only technology or the allocation of resources must change, but also institutions. He (1994: 4-5) has argued that, for the analysis in developing countries, where "modern" institutions (such as markets) are not fully developed, new institutional economics seems more useful and suitable than neoclassical economics, which treats institutions as a given (or overstresses economic factors for institutional change).

Although new institutional economists also focus on resource endowment, product demand, relative prices and technology as the main reasons for institutional change, they acknowledge the importance of non-economic factors such as power, ideology, religion, tradition or values. In particular, *new institutional economic anthropology* has suggested a mutual influence between economic factors such as relative prices and "ideology," that is, cultural values and beliefs (Ensminger 1992). Often, cultural values and beliefs change in order to conform with changed economic, social and political structures. However, cultural values and beliefs are "lumpy"; i.e., they are relatively invulnerable to change. Only considerable and sustained changes in relative prices, for example, can alter cultural values and beliefs. In turn, changing cultural values and beliefs also influence relative prices (Ensminger 1992: 10-11). In neoclassical economics, by contrast, values and beliefs (referred to as individual tastes or preferences) were falsely assumed to be constant.

Furthermore, consideration of the "cultural endowment" or the "ideology" is crucial for development interventions that aim to introduce new institutions such as market institutions, or new state regulations and support systems. "Modern" institutions such as market institutions or particular state laws that work well in Western countries (and are therefore advocated in structural-adjustment programs for developing countries) may not produce the expected results in other societies: imported institutions may not be compatible with the cultural endowment or with particular *indigenous institutions*. Indigenous institutions – He (1994) called them *micro institutions*, de Janvry *et al.* (1993) referred to them as *civil society* – may encompass many different forms of reciprocity and cooperation among kin groups or people of the same village, forms of patronage, etc. Informal and formal indigenous institutions are often the key to the success or failure of planned institutional innovation. According to new institutional economics, the reasons for the development success of newly industrialized countries such as Taiwan and South Korea, for example, cannot be found in market liberalization or in specific, selected state interventions alone.

Rather, this development success was built on flexible indigenous institutions such as family networks that supported the functioning of markets and that facilitated the operation of government policies (He 1994: 27).

However, in an environment characterized by pervasive risks and asymmetry of information and power, neither "market forces" (i.e., abstract forces of supply and demand) nor the state are efficient. In other words, both market failure and state failure may occur. In such situations, not only may traditional non-market institutions persist (Stiglitz 1986; Popkin 1988) but, also, new *community-based organizations* may emerge to fill the vacuum created when the state withdraws its support. Community-based organizations are based on common interests, consensus and shared values that lead voluntary collective action. They include social networks, grassroots organizations, social movements, self-help groups and cooperatives.

The emergence of community-based organizations is also facilitated by indigenous institutions (i.e., the existence of traditional forms of cooperation or reciprocity) and by the support of development organizations. Non-governmental development organizations have been particularly inclined to recognize the potential of local communities to organize themselves and apply their knowledge and skills (Farrington *et al.* 1993). Non-governmental organizations have therefore become advocates of *community-based development strategies* that would develop and use this potential (see Section 2.5). However, political, socioeconomic and sociocultural conditions may constrain self-help and participation so that local groups rely heavily on the support of non-governmental organizations. This, in turn, might sometimes lead not to empowerment, but to a new dependence on an outside organization (Véron 1993: 124-126).

Community-based organizations must also be seen in their relation to the state (which may either repress or encourage a particular community-based organization) and in relation to markets (of which a cultivators' organization may be able to make more effective use than individual cultivators on their own). New marketing opportunities and the integration in a wider economy may not only reinforce but also alter exploitative power structures. Furthermore, potential complementarities between community-based organizations, markets and the state need to be taken into consideration, both by development theory and in practice, thus transcending the old dichotomy of market-led versus state-led development models, or of market liberalism versus state interventionism (He 1994: 23-24).

3.6 Summary

The neoclassical, neo-Marxist and new institutional theories I have discussed contribute to a better understanding of the relationships between markets, agricultural change and socioeconomic development. These theories indicate that crop markets can influence agricultural practice in form of:
- relative prices, supply-and-demand structures and transaction costs (structural properties or general market conditions);
- particular marketing arrangements, contractual interlocking and power relations (vertical relations between buyer and seller);
- relations to production and state regulation, marketing technology and marketing infrastructure, social norms and cultural values (wider context in which crop markets are embedded).

However, generalizations about the relation between crop markets and agricultural practice are difficult to make because agricultural markets are highly diverse and complex:

> Clearly, little can be assumed about agricultural markets. The student has to be braced for the possibility of extreme diversity and complexity in the structure, institutions and behaviour of marketing system. While an immense effort has been put into the description, theorisation and modelling of stylised forms of agricultural production, the institutional complexity of actual markets has been neglected. Marketing is far from being one layer of transaction between producer and consumer. "Marketing" is a system of transactions and transfers of property rights over commodities, of trading and production activities, at any point in which structural elements, power relations and contractual forms may vary. (Harriss-White 1996a: 49-50)

Furthermore, this theoretical discussion indicates that not only markets but also interlinked factors such as production technology, agricultural infrastructure and development interventions influence agricultural change.

Depending on the applied concept of market exchange, development theories have drawn contrary conclusions about the contribution of markets to agricultural development: "Free" trade and the utilization of *comparative advantage* lead to agricultural growth; *forced commerce* results in the exploitation of poor peasants; *transaction costs* determine the developmental outcome of trade. Combining these theories with the approaches to sustainable development (see Ch. 2), one can also identify three chains of arguments (neoclassical-economist, Marxist-influenced and institutional argumentations). This summary highlights the differences between these viewpoints with respect to the influence of crop markets on agricultural practice and their consequences for sustainable development. No attempt will be made to do justice to particular authors, but rather to represent ideal types.

The *neoclassical-economist viewpoint* concentrates on economic factors to explain agricultural and environmental change, and to suggest policies for sustainable development. Neoclassical theory has regarded individuals as the prime movers of development. The theory holds that farmers and other agents react directly to changes in relative prices. The sum of these "rational" individual decisions (i.e., the force of demand and supply) in combination with the utilization of *comparative advantage* through trade and specialization would bring about economically "optimal" cropping patterns and thus economic development. In developing countries, however, commercialization and markets are not fully developed so that farmers may not realize the potential benefits of trade. According to neoclassical theory, it is therefore imperative to develop agricultural markets so that farmers can respond "rationally" to price incentives and make use of their comparative advantage.

Furthermore, the neoclassical-economist viewpoint suggests that both underdevelopment and environmental degradation originate from the lack of well-developed markets. The environment may be overused because it is underpriced or has no price at all. Because the environment has public-goods character, moreover, *markets for the environment* do not emerge automatically, but must be created by state intervention. Once the environment is integrated into the market system (e.g., through environmental taxation, tradable resource-use and pollution certificates), environmental goods and services would form significant input costs for (agricultural) production. Moreover, if natural resources were given the "right" price, the environment would not be overused but allocated efficiently among other factors of production. Because of comparative advantage, furthermore, farmers would grow particular crops where it is economically *and* ecologically the most efficient to do so, at least, as long as the gains from trade are higher than transportation costs.

In summary, the neoclassical-economist viewpoint suggests that agricultural practice (i.e., changes in cropping pattern, in cultivation intensity and in cultivation methods) would be directly dependent on changes in relative prices. Crop markets (like other agricultural markets) give price signals to which farmers respond. Furthermore, crop markets are believed to further the utilization of comparative advantage. Agricultural marketing, according to neoclassical theory, is therefore beneficial to socioeconomic development; but markets may have positive or negative effects on the environment, depending on whether or not prices reflect environmental costs.

By contrast, the *Marxist-influenced viewpoint* takes into account the wider economic, social and political context, and is more critical to the effects of crop markets. Neo-Marxist theory suggests that trade does not lead to comparative advantage for peasants because of the unequal terms on which different classes are involved in market exchange. Powerful classes may shape market conditions to their own benefit and interlock crop markets with credit, land and labor. Eventually, peasants can be compulsively involved and exploited in agricultural markets. In this way, markets induce changes in cropping pattern and in cultivation methods and shape agricultural production. Commercialization makes peasants vulnerable to market fluctuations and subject to the *simple reproduction squeeze.*

Moreover, political ecology suggests that the simple reproduction squeeze and *unequal access to resources* not only perpetuate social injustice but also lead to environmental degradation. Sociopolitical and economic dependence and exploitation, which are often enhanced by commercialization under prevailing agrarian relations, compel many poor peasants in the Third World to overuse natural resources and thereby degrade the environment. Peasants may also be driven to grow cash crops, the cultivation of which is often not suited to local bio-physical conditions. At the same time, unequal access to resources is responsible for over-consumption by the rich and for environmentally unsound, resource-depleting cultivation by capitalist farmers or enterprises. These may follow short-term profit-maximizing strategies because they can easily move on with their profits once the land has become degraded. In order to achieve sustainable development, therefore, fundamental, if not revolutionary, social change may appear to be necessary. Yet, such reformist measures as land reforms, rural-credit schemes and state-controlled marketing may also be able to mitigate the dependence and exploitation of the poor.

To sum up, the Marxist-influenced viewpoint suggests that agricultural practice is influenced by crop markets that are part of the totality of social, political and economic agrarian relations. Because dominant classes control agrarian relations, however, trade is not beneficial to poor peasants and crop markets become a means of exploitation. The exploitation of poor peasants through markets also leads to the exploitation of nature and to unsustainable development.

The *institutional viewpoint* generally gives a less definite scenario. It also draws the attention to local conditions and practices as well as to indigenous institutions and knowledge. As presented here, the institutional viewpoint combines new institutional theory with the indigenous-knowledge approach but, unlike community-based development, assumes that outside intervention and market involvement may also have a positive effect on sustainable development.

New institutional economics has considered a multitude of factors (e.g., relative prices, indigenous and modern institutions, power relations and cultural values) that can influence cropping patterns and cultivation methods. It maintains that crop markets influence farmers' practices not only in terms of relative prices but also through their infrastructural and institutional aspects. Only if *exchange costs* are kept low, will trade in agricultural products be beneficial for farmers. Equal distribution of potential gains from trade, moreover, depends on whether exploitative power relations can be broken or bypassed. Furthermore, development not only relies on the market or the state, but also on *indigenous institutions* and their compatibility with "imported," "modern" institutions.

Indigenous knowledge can be interpreted as that facet of indigenous institutions of particular relevance to sustainable development. This is because indigenous resource-use systems are often well-adapted to local, bio-physical and socioeconomic, conditions. Community-based development strategies have identified great development potential of "local" people, indigenous technical knowledge and community-based organizations. The institutional viewpoint maintains that community-based organizations, but also market-based instruments, tailor-made state interventions and democratic institutions, can be useful tools to achieve sustainable development. However, if it is to their particularistic benefit, powerful interest groups may hinder technological, infrastructural and institutional development or even advocate unsustainable practices.

In summary, the institutional viewpoint suggests that agricultural practice is influenced by many dimensions of crop markets. Agricultural marketing is neither necessarily beneficial nor inevitably detrimental to socioeconomic development. Environmental consequences are also undetermined. Specialization resulting from trade, for example, may both reduce local biodiversity and enable cropping patterns that are in accordance with (ecological) comparative advantage. Market-oriented cultivation may both stimulate the application of appropriate new technologies and supersede useful indigenous technical knowledge and environmentally sound practices.

This theoretical discussion has not provided conclusive evidence about the relation between crop markets, agricultural change and sustainable development. Yet, the conflicting hypotheses of the neoclassical-economist, Marxist-influenced and institutional viewpoints are useful heuristic tools to supplement the concepts of social practice and real markets. The following chapters, which deal with sustainable development and agrarian relations *in practice*, will frequently refer back to hypotheses generated by the theories discussed in this and the previous chapter.

4 From Theory to Field Research

Empirical research depends on assumptions made and on the applied theoretical frameworks and viewpoints. In turn, empirical data may call into question the assumptions and hypotheses of particular theories. In this study, however, the various theoretical viewpoints are not tested in a strict sense. That is, the research was not designed in order to verify or falsify particular predetermined hypotheses derived from any specific theory. Rather, the theoretical viewpoints and the concepts of social practice and real markets have provided general guidelines (as opposed to rigid tools) for data collection, analysis and interpretation. Definitions, classifications and operationalizations, as well as the analytical framework and methods used for this study, have been adapted to the context and the research topic. Furthermore, they have been continuously modified according to experiences during the field research. They are as much a product of the research process as of theoretical considerations.

4.1 From Theoretical Concepts to Empirical Concepts

This study focuses on *agricultural practice*, its *socioeconomic and technical-material conditions* (including crop markets, labor and land relations, technology, bio-physical conditions, etc.) and its socioeconomic and environmental consequences that have implications for *sustainable development*. For empirical research, it is necessary to clarify and operationalize these key terms. In doing so, I have used as sensitizing devices the concepts of social practice and real markets, together with theories of the "middle range" regarding cultivators, agricultural change and sustainable development.

Sustainable development has already been defined as "development that meets the needs of the present without compromising the ability of future generations to meet their own needs" (see Section 2.2). This concept, however, requires proper operationalization to become useful for empirical research. This study looks at sustainable development through the positive and negative effects of actual practices and processes on socioeconomic development and the environment. Thereby, the environment is primarily viewed as a determinant of general human well-being in the present and in the future.

The socioeconomic and environmental consequences of an individual farmer's practices form new conditions for the farm household, for neighboring farmers, for other people in the community, for contemporaries who live far away, and even for future generations. Impacts on the livelihoods of cultivating households may include generation of wealth, income diversification, self-employment, capability to cope with changing socioeconomic or bio-physical conditions, and enhancement or degradation of bio-physical resources. Impacts on non-cultivating households include mainly employment opportunities in agriculture-related jobs and food security. In general, however, one must bear in mind that agriculture is only one of many interrelated areas that may contribute to sustainable or un-sustainable development, and environmental regeneration or degradation.

Agricultural practice (or *cultivation)* stands for the practice of growing particular crops. It involves activities and processes regarding cropping patterns (the combination of crops that are grown), cultivation intensity and cultivation methods (e.g., monoculture, intercropping, irrigation, fertilizer use). Usually, *cultivators* are the principal agents of agricultural practice. According to Webster's Ninth Collegiate Dictionary (1991), a *cultivator* is very generally "one that cultivates"; a *farmer* is "a person who cultivates land or crops"; a *peasant* is "a member of a ... class of persons tilling the soil as small landowners or as laborers" or "a usually uneducated person." Although the words cultivator, grower, farmer and peasant have different connotations, they are used as synonyms in this study.

In Kerala, the cultivators are usually landowners because all forms of tenancy were abolished in 1970. Kerala's common definition of cultivators also includes those landowners who do not till the land themselves but only supervise and control the work on the field. Still, these *supervising cultivators* are the principal decision-makers regarding cultivation. In fact, many farmers in Kerala (even smallholders) hire agricultural laborers for most work on the field. So, the cultivators are often not the actual *tillers* of the soil.

In order to avoid the misconception that farmers form a homogeneous class, it is useful to divide cultivators into categories. The empirical research has confirmed that particular farming practices often correlate with the socioeconomic position of a farmer's household. However, the conventional farmer classifications that are based on a few particular aspects of the household's socioeconomic position did not provide adequate categories for this study.

India's common governmental classification, for example, has categorized farmers according to the size of their landholdings, into marginal, small, semi-medium, medium and large farmers. However, landholding size is not a good indicator of socioeconomic position. The category "marginal farmers," for example, includes not only poor agricultural workers with a small plot of land, but also relatively rich government employees, professionals or businessmen who grow a few crops in their home garden. Moreover, India's common classification is not adapted to the small-holding structure of Kerala, where more than 90 percent fall into only one category; that of marginal farmers.

Marxist peasant theories that build upon the relations of production provide more meaningful cultivator classifications. Landless agricultural laborers are defined as a category of people that are forced to regularly sell their own labor-power. Poor peasants are also forced to sell their labor but they retain a small plot for cultivation. Middle peasants or family farmers are able to rely fully on the cultivation of their own land with the use of family labor. Rich farmers not only use family labor but also hire labor, and they use "modern" means of production (e.g., capital, agricultural machinery). Capitalist farmers, finally, are like rich farmers but they fully rely on hired labor (Roemer 1982, cited in Jose 1991: 92-93; Bernstein 1982: 170). In the Kerala context, however, this classification is inappropriate because most cultivators (including cultivators who are not rich at all) employ wage laborers. Moreover, the relations of production *within* agriculture do not reflect the general socioeconomic position of a household because off-farm employment is excluded. For example, a not very wealthy minor bank clerk who retains a plot for cultivation and hires agricultural labor would be labeled "rich farmer" – a term that is obviously misleading for such a person.

My own farmer classification, on the other hand, is not based on single indicators such as monetary income, landholding size or labor hiring but on the entirety of tangible and intangible assets of the farmer's household. Tangible assets include stores and resources such as land, labor, capital, income, employment, information, education, technology, buildings, food. Intangible assets include actual access to resources as well as claims that can be made from the family, the community or the state (see Chambers & Conway 1992). Moreover, *negative* claims (obligations) need to be considered. In the Kerala context, the dowry forms a very substantial claim or obligation, respectively.

The entire assets of a household are appraised qualitatively through the informants' own assessment supplemented with the judgement of neighbors and of my interpreter. On a continuum, farmers are categorized into poor, (lower- and upper-) middle class and rich. These categories are relative and valid only in the local context. People classified as lower-middle class farmers in this study would, in national or international surveys, probably be added to the population living below the poverty line.

In addition to their socioeconomic position, the farmers are also classified according to the relative importance of cultivation to the respective household. Part-time farming is widespread in Kerala (see Section 5.4.). Households, and even single household members, often combine various sources of income. For the purpose of this study, a distinction between those farmers for whom cultivation is the most important occupation (full-time farmers) and those for whom cultivation only forms a secondary occupation (part-time farmers) is made. If not specified, the term "farmer" relates to both full-time and part-time farmers.

In Kerala, poor and middle-class part-time farmers are the most common. In terms of area, however, rich and middle-class farmers are the most significant. In brief, the most important types of cultivators include:
- poor part-time farmers (e.g., wage laborers who grow a few crops on the homestead);
- *poor full-time farmers* (e.g., resource-poor smallholders with a clear-cut interest in agriculture);
- *middle-class part-time farmers* (e.g., lower-rank employees or businesspeople who retain agricultural land);
- *middle-class full-time farmers* (e.g., family farmers or middle peasants with a clear-cut interest in agriculture);
- *rich part-time farmers* (e.g., higher-rank employees or businesspeople who retain agricultural land);
- *rich full-time farmers* (e.g., big landowners with a main interest in agriculture).

The focus on households (mostly nuclear families) seems justified because they are the main micro-level socioeconomic units in Kerala. However, within a household, (agricultural) decision-making and work are highly differentiated according to gender and age, and tend to be very complex. Adequate attention to these issues would require a full, separate study.

While the socioeconomic position of individual households may account for the variation from cultivator to cultivator, overall agricultural processes are better explained by general conditions for agricultural practice, *socioeconomic* and *technical-material* factors. This study focuses on a number of interrelated factors that are often used in studies on agriculture and – since theories and theoretical categories tend to recur to agents – also by the cultivators themselves. For different types of cultivators, these factors may form either opportunities or constraints, and may be altered partly through the cultivators' practices.

Socioeconomic factors *(agrarian relations)* are divided into:
- *farm-output relations*, particularly *crop markets* (e.g., individual market access, personal and social relations to buyers, contractual interlocking and marketing arrangements; importance of home consumption and reciprocity *versus* sale; relative prices, transaction costs, degree of competition, seasonal and spatial demand-and-supply structures; market infrastructure, state regulations and social norms regarding marketing);
- *capital/agricultural-input relations* (e.g., access to credit, seeds, fertilizers and irrigation; availability, terms and conditions of formal and informal loans);
- *labor relations* (e.g., hired labor *versus* family labor; social relations to workers, influence of labor unions, local wage rate, seasonal labor supply);
- *land relations* (e.g., property-right regimes, leasing arrangements, land markets).

Technical-material factors include:
- *technology* (e.g., information, access, local suitability);
- *agricultural infrastructure* (e.g., availability and access to irrigation);
- *bio-physical factors* (e.g., soil, climate, vegetation).

This study focuses on the relative significance of crop markets or agricultural marketing as a condition for sustainable or unsustainable agricultural practice. Agricultural marketing includes interactions ranging from the cultivators' sale of agricultural commodities to the consumption of the raw or processed products; and it also involves productive activities (e.g., assembly, post-harvest treatment, storage, processing, transport).

4.2 Wider Contexts and Localized Practices

The empirical research breaks into two parts: a macro-level appraisal of development, environmental and agricultural change, socioeconomic and technical-material factors for farming in Kerala; and two case studies on the cultivation and marketing of particular crops in selected localities. The focus is on cultivators, agricultural practice and repeated localized inter-actions in the fields of agricultural marketing, credit, labor, land and techno-logy. However, localized practices and interactions are not comprehensible without consideration of the wider contexts, in this case the regional and crop-specific conditions for cultivating pineapple or cashew.

Although the linear structure of this book would suggest otherwise, the wider contexts and the localized practices of cultivators were not studied in a linear sequence but rather in a circular fashion. Contexts and localized practices are mutually dependent. For example, cultivators point out the relevant conditions for farming but they may have only incomplete knowledge about these conditions (unacknowledged conditions). Therefore, socioeconomic and technical-material factors of farming have to be studied on other levels too (e.g., through information from traders, consumers, government officers, secondary literature). On the other hand, knowledge about the broader context of the farm sector without comprehension of the cultivators' perspective is insufficient to explain changes in agricultural practice. Only by studying local social practice in conjunction with the explanations of the knowledgeable cultivators can one determine the influence of changing agrarian relations, of technological and infrastructural changes – and thus, also the impact of planned development interventions.

For data collection, analysis and interpretation, therefore, the entry point is the cultivators' agricultural practice as part of the household's entire agricultural income and livelihood. Attention is paid to different practices, reasons and knowledge among the various types of cultivators. Furthermore, changes in cropping patterns and cultivation methods are analyzed, beginning with the cultivators' own explanations, which hint at the relevant socioeconomic and technical-material factors. Together with material from other sources, information on the cultivators' perspective allows one to interpret the relative importance of the various factors, with special reference to crop markets. Crop markets, access to capital, agricul-tural inputs and technology, labor and land relations are discussed as repeated, institutionalized interactions between cultivators and other agents (e.g., traders, agricultural officers, wage laborers, neighbors) and as structural properties of the related institutions. Special attention is paid to the reproduction and transformation of institutions that influence agricul-tural practice. As far as possible, the motives and values of farmers that lie behind their expressed reasons are also examined.

Furthermore, the environmental and socioeconomic consequences of changes in cropping patterns and cultivation methods are assessed in a qualitative way, to estimate their impact on cultivators and other economic groups, on neighbors and people living elsewhere, on contemporaries and future generations. Sustainable development is thus addressed in a human-centered way that considers the particular socioeconomic position of a household but is not limited to a spatial unit or to agriculture alone. In particular, the relative influence of crop markets on either sustainable or unsustainable agricultural practices, and therefore their relative positive or negative contribution to sustainable development, are appraised. In order to judge environmental awareness (or the lack of it) and the society-specific interpretation of sustainability, people's own perceptions regarding the problematic environmental and socioeconomic changes are also considered.

4.3 Case Studies: Pineapple and Cashew

The region chosen for this study (Kerala) is exceptional in many respects. Particularly in terms of social development, Kerala is very "advanced." As regards the conventional indicators of social development, for instance, Kerala has a low infant-mortality rate, a life expectancy of 72 years and a literacy rate of nearly 100 percent. Recently, Kerala has also been cited as a positive example of environmental protection and of sustainable development:

> [Kerala is] a paradigmatic case of a state or society that has some characteristic features of what might be deemed sustainable development. (Parayil 1996: 941)

The backwardness of industrial development in Kerala in part accounts for the fact that the environmental situation there is comparatively good, despite the fact that Kerala's 29 million people are confined to an area equal to that of Switzerland. However, Kerala is certainly not void of environmental problems: decline of wetland ecosystems; loss of biodiversity; deforestation; soil erosion; chemical pollution of soils and water (see Section 5.3).

In the case studies of pineapple cultivation in the region around the village of Vazhakulam, Ernakulam District, and of cashew cultivation in the region around the villages of Mattanur and Iritty, Kannur District, I have focused on the description and analysis of localized practices and processes in the period between 1980 and 1995. However, particularly in the descriptive sections, I have also taken the broader context into account, and included information about pineapple and cashew cultivation in former times.

Furthermore, these selected crops are not studied in isolation but in relation to the whole farming system and the household economy. The selected case studies may not be representative but, at least, they indicate some general processes of Kerala's agriculture, such as the conversion of wetland paddies or the spread of rubber cultivation.

The cultivation of pineapple and of cashew differ in several important respects. Pineapple is a highly perishable crop grown on irrigated and rain-fed dryland as well as on converted wetland. Cashew, on the other hand, is a moderately perishable tree crop grown only on rain-fed dryland. Moreover, these case studies shed light on different facets of the environmental problem in Kerala. Pineapple cultivation, which is expanding rapidly and is being intensified in Vazhakulam, relates to common environmental problems such as the widespread conversion of wetland paddies into dryland or the general "chemicization" of agriculture. On the other hand, cashew cultivation tends to be environmentally sustainable and even to be effective for soil conservation. Yet, rubber monoculture is partly replacing old cashew plantations in some areas of Kerala, including Mattanur-Iritty.

Cashew and pineapple are also marketed differently. In Kerala, which has boasted a substantial export-oriented agricultural sector since the 19th century, long-established markets for "traditional" cash crops (e.g., cashew, spices, rubber) coexist with newly emerging markets for such particular crops as fruit, vegetables, medicinal plants and flowers. Marketing opportunities for raw cashewnut are almost the same everywhere in Kerala. Because of the export orientation of Kerala's cashew-processing industry, local raw-cashewnut markets are linked to the global market for both raw cashewnut and processed cashew kernel. As well, there has been massive government intervention in the cultivation, processing and marketing of cashew. For instance, Kerala's cashew market has undergone gradual but constant change since India's government liberalized cashew imports and exports in the 1980s, and Kerala's government deregulated the raw-cashewnut market in 1995.

Pineapple, on the other hand, is an example of a new cash crop produced for the domestic market. In India, market demand for pineapple is increasing because of the rapidly growing urban middle class that can afford to buy relatively expensive fruit in the market and has little scope to grow fruit on its own. For pineapple, however, markets and market opportunities are regionally concentrated. Furthermore, pineapple cultivation and marketing have received little government support in Kerala, and export opportunities for pineapple have hardly been explored.

4.4 Field Methods

The study relies on primary data collected during a 15-month field visit in Kerala from August 1994 to November 1995 and a revisit in October 1996. It also draws on secondary data from official statistics, government reports, newspapers, journals, published books and the internet. For primary data collection, a plurality of *qualitative* social science methods were applied, including interviews, tools from "Rapid Rural Appraisals" and observation. Qualitative methods are more appropriate than quantitative methods to explore the cultivators' reasons for particular practices. I therefore used qualitative methods primarily to discover *why* cultivators apply sustainable or unsustainable practices, although I was also able to derive some quantitative estimates from the qualitative data (for a full justification of the use of qualitative methods see Flick (1995)). For the broad picture, however, quantitative data from secondary sources proved essential. Regarding environmental issues (e.g., the environmental consequences of pineapple or cashew cultivation), this study relies on information from scientists, documented experiences in other areas, and the researcher's observations. Due to logistic constraints, scientific field or laboratory experiments were not carried out. However, the study did justice to the social aspect of environmental issues by assessing and interpreting how local people perceived the "problems" related to pineapple or cashew cultivation, respectively.

Primary data collection took place in two main, and several secondary, research sites in Kerala. For the case study on pineapple cultivation, research concentrated on the area around the village of Vazhakulam, Ernakulam District (Central Kerala). The case study on cashew cultivation is placed in the rural area of Mattanur-Iritty, Kannur District (North Kerala). Starting from the city of Thiruvananthapuram (South Kerala), five week-long field trips in different seasons took me to each of these research sites. For comparison with the situation in the main research sites, and to gain additional insights into cultivators' livelihoods and agricultural trends in Kerala, research was also carried out in five secondary research sites in Thiruvananthapuram, Pathanamthitta, Thrissur, Palakkad and Kozhikode Districts, each of which was visited once or twice. In addition, the study included visits to 10 agro-processing factories in four towns (Kollam, Kottayam, Thrissur, Kannur) and investigations in two important urban fruit markets (Thiruvananthapuram, Kochi-Ernakulam). Finally, about 30 expert interviews were carried out with scientists, agronomists, marketing specialists, development officials and NGO-workers.

In total, nearly 100 farm households, differing in socioeconomic status, were visited. Care was taken to include both farm households near main roads and those in the interior. I also interviewed some key informants, identified with the help of local civil servants, who had special knowledge about the history of the place, particular agricultural practices, and environmental issues. Apart from farm visits, about 20 semi-structured, individual or group interviews each with wage laborers, traders and local officials were carried out in the main research sites. Observation and such tools of Rapid Rural Appraisals as transects, social mapping, and time lines were used to gain an introduction to the locality and its history, and to make a quick assessment of environmental changes.

The nearly 100 farm visits, most of which lasted between half-an-hour and one hour, form the cornerstone of the primary data collection. *Semi-structured in-depth interviews* with farmers produced factual information about their livelihoods, agricultural practices, marketing arrangements, access to technology and local soil conditions. Furthermore, the semi-structured interviews provided crucial information about the farmers' reasons for particular practices and their perceptions of general developments, particularly environmental and socioeconomic changes in the village. As opposed to totally unstructured interviews, semi-structured interviews with a checklist helped to ensure that the core issues were addressed and that the interviews were comparable. As opposed to standardized questionnaire surveys, on the other hand, the semi-structured interviews helped to create a fairly relaxed and natural atmosphere; and the flexible use of the checklist according to the particular knowledge of the informant made it possible to explore crucial issues in depth. As well, the semi-structured interviews produced unexpected insights and drew attention to emerging issues that proved important to the study. Using this method made it possible during my time in the field continuously to revise both the checklist and the research hypotheses.

During the farm visits, *observation* of the informants' plots (that often were far away from their dwellings) served to verify and clarify oral information. This triangulation was necessary to distinguish between the cultivator's actual practices and his or her personal situation, on the one hand, and perceptions of general trends in the village, on the other hand. Graphic tools used in Rapid Rural Appraisals were also applied. The ranking method was used to get to know about cultivators' preferences and motives, and seasonal calendars were used to learn about seasonal workload, prices and income from various crops and wage labor. However, in Kerala, where most farmers are formally educated, these graphic tools appeared redundant to the informants. Generally, the cultivators were able and willing to give the information verbally. Notes were taken during the farm visits, unless the informant was initially frightened or suspicious. Casual encounters with cultivators in government offices, the marketplace, local restaurants, tea and toddy shops or on the bus provided informative insights that supplemented what was learned during the farm visits.

The main difficulty in carrying out research with qualitative methods was language. For most interviews, I depended on an interpreter from Kerala. Under these circumstances, it is sometimes difficult to create an informal atmosphere and to react quickly to newly emerging issues. Moreover, Kerala's farmers are unaccustomed to seemingly unstructured conversations with researchers. Surprisingly, they are more familiar with questionnaire surveys. Nevertheless, the farmers were open and very cooperative. Traders, on the other hand, were more reluctant to disclose information, which is often treated as a business secret. Much time is thus needed to gain the traders' confidence and collaboration. Furthermore, as a male researcher in Kerala, it is difficult to arrange interviews with women (e.g., female members of cultivating households) and to get their views adequately represented. Generally, this study does therefore not deal with intra-household and intra-family relations: to study these effectively would require more systematic application of methods such as participant observation. For the purpose of this study, however, the employed methods were useful.

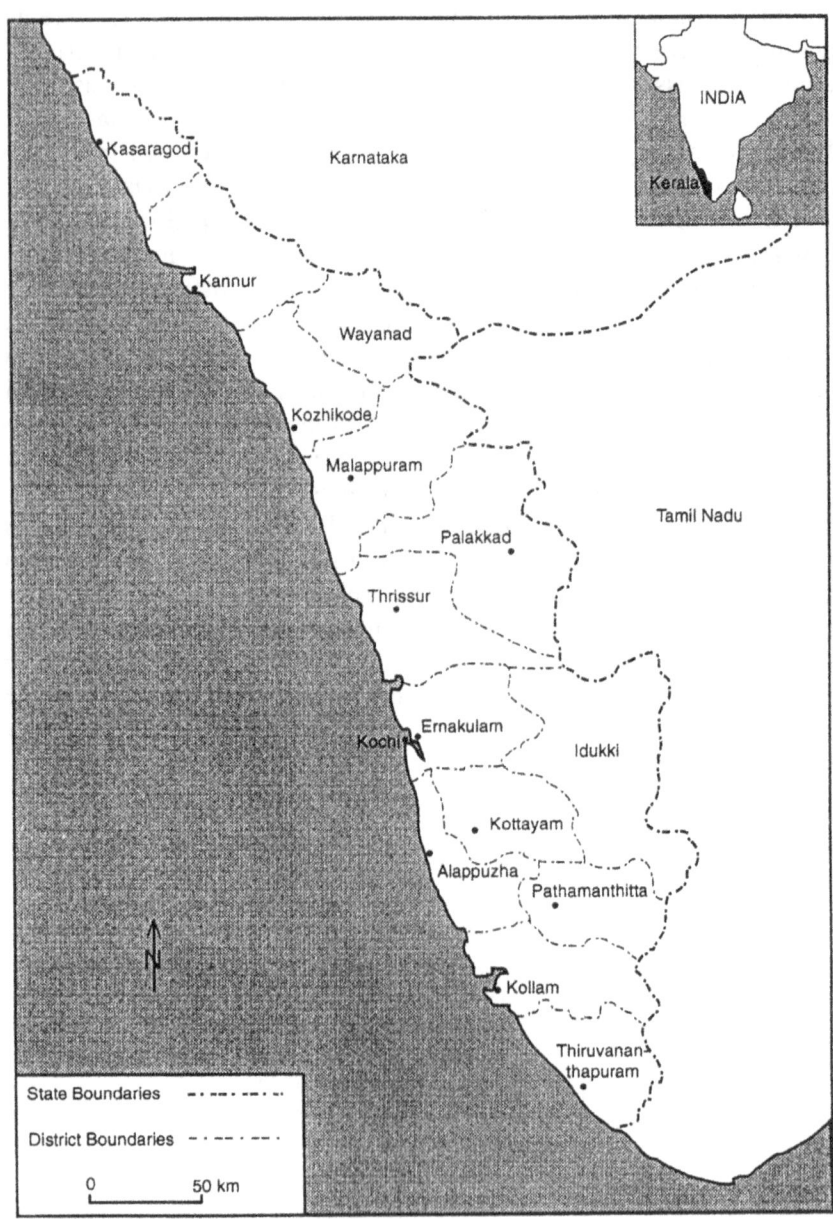

Fig. 3 Map of Kerala
Source: Basic Map after Nossiter (1988).

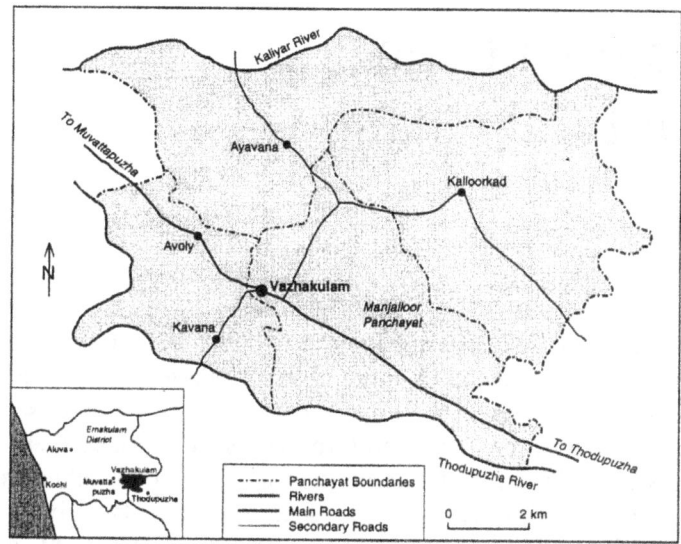

Fig. 4   Map of Vazhakulam
Source: Census of India; my own investigations.

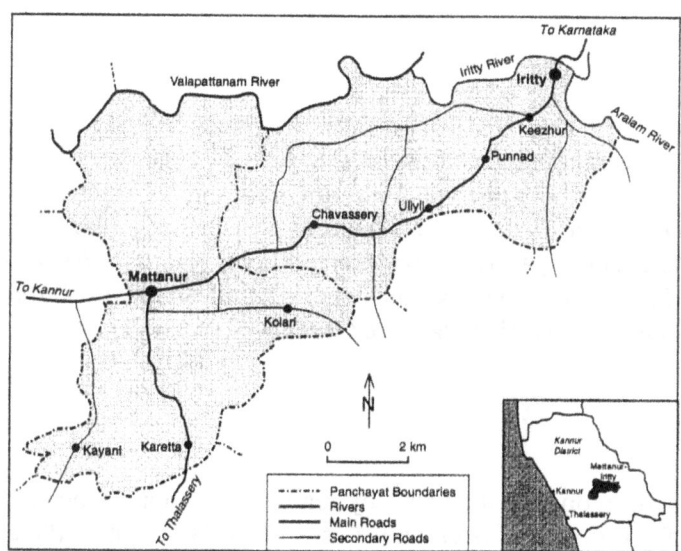

Fig. 5   Map of Mattanur-Iritty
Source: Census of India; my own investigations.

5 Kerala: Development, Environment and Agricultural Change

In today's world, people's lives are not limited to their daily face-to-face interactions. Rather, they are shaped by more abstract relations in a broader social and political system, and in a globalized economy. To understand people's practices and decisions, it is therefore necessary to place them in this broader context. While parts of this context are universal, others are specific to particular societies, regions and social groups. By giving a general overview of Kerala's society, development and environment, and by discussing socioeconomic and technical-material conditions for agriculture in Kerala, this chapter attempts to elucidate the context that influences and structures the decisions and practices of cultivators in that part of South India.

5.1 State Profile

The Indian state of Kerala is located in the south-west of the subcontinent, on a narrow strip of land between the Western Ghats and the Arabian Sea. This land may have been named after *kera-lam*, Sanskrit for "the land of coconut," because coconut palms are widespread here; or after *Chera-lam*, Sanskrit for "the land of the Chera Kings" who ruled in Southwest India in the 9th and 10th century A.D. (Kerala State Gazetteer 1986: 6-7).

5.1.1 History

Kerala's coast has been open to the world by sea for a very long time. Trade in spices with Babylon may go back as far as to 3000 B.C. Later, Phoenician, Greek, Roman, Arab, Jewish and Chinese traders did business in Kerala, which became an important entrepot connecting South Asia with

Europe, the Middle East and China. It is often mentioned that Kerala's particular geographical location – isolated from the east and the rest of India by the Western Ghats and open to the Arabian sea and, therefore, to international seafarers – contributed to the early emergence of international trade and a general outward orientation of its people. Politically, during most of the pre-colonial period, Kerala was split into numerous local kingdoms that developed and vanished, conquering and reconquering each other's territory. By the time the Portuguese arrived, the warrior aristocracy and the land-owning clergy were as powerful as the monarchies (K.V. Eapen 1986).

In 1498, the Portuguese navigator Vasco da Gama landed in Calicut (North Kerala), marking the onset of the colonial era, which would last for almost 450 years. Unlike earlier foreign traders, the Portuguese, and later the Dutch, French and British, too, sought monopoly rights in the spice trade and used not only diplomacy but also military power against local rulers. In 1792, the British succeeded in establishing hegemonial power in Kerala. The *British rule*, which lasted for more than 150 years, went beyond mere protection of colonial trading interests to political and administrative control of the territory. Malabar (North Kerala) came under direct British rule as a district of Madras Presidency; the *princely states* of Cochin and Travancore (central and south Kerala, respectively) and their *rajas* (Hindi for kings) came under indirect British rule. Eventually, the princely states developed in another way than Malabar district. The Rajas of Travancore, in particular, initiated social reforms and invested a great deal in education and infrastructure. In Malabar, on the other hand, British law strengthened the feudal system. Furthermore, land taxes there were very high. In the 19th century (especially after the opening of the Suez canal in 1869), the plantation economy (pepper, cardamom, coffee, tea, rubber) started on a large scale in the highland of both Malabar and the princely states (K.V. Eapen 1986; Kurien 1994a).

In the first half of the 20th century, Kerala's society underwent tremendous change. Values were influenced by formal education, exposure to Western liberal ideas and the work of missionaries. Moreover, caste-based associations and movements mobilized the masses for far-reaching reforms that weakened traditional institutions such as the rigid caste structure and the joint-family system. From the 1930s onward, socialist, communist and trade-union movements have also engaged in the struggle for a more equal society. Kerala's social transformation in the 20th century followed many patterns of social "modernization," shifting, for example, from inherited status to achieved status, from interdependence of caste to competition of individuals, and from traditional authority to modern bureaucracy (Jeffrey 1976, cited in Saradamoni 1983: 265).

Notable political change came with the termination of the British *raj* (Hindi for rule) and the Independence of India in 1947. In the course of the formation of new states according to linguistic boundaries in 1956, the *Malayalam*-speaking regions of Travancore, Cochin and Malabar (and South Canara, a small area north of Malabar) were unified as the state of Kerala, one of 25 states of the Indian Union.

5.1.2 Demographic and Administrative Features

Kerala is a comparatively small state, covering an area of approximately 39,000 square kilometers (1.5 percent of India). According to the 1991 Census, Kerala has about 29 million inhabitants (3.5 percent of India's total population of approximately 850 million). The population density is among the highest in the world: 747 persons per square kilometer, as compared with 267 in India as a whole (or 123 in China; 27 in the United States). 81 percent of Kerala's people live in rural areas (as compared with 74 percent in India). The urban population is growing at a relatively low rate. However, in the case of Kerala, the distinction between urban and rural areas is problematic because of the unique settlement pattern. In rural Kerala, the population does not live in clustered villages but on scattered homesteads. These areas are populated very densely and their boundaries with the few interspersed small towns are blurry. The unique settlement pattern of Kerala is appropriately called *rurban* (Pronk 1997). Moreover, infrastructure and lifestyles in Kerala's rurban and urban areas are similar, and social indicators of development are only slightly different. Urbanization (or *rurbanization)* takes place less in the cities than alongside state and national highways and at intersections of country roads.

Kerala is divided into 14 *districts* (see Fig. 3), 151 *community-development blocks*, nearly 1,000 *panchayats* (local political-administrative authorities in rural areas), about 60 *municipalities* (town areas) and three *corporations* (the cities of Kochi (Cochin), Thiruvananthapuram and Kozhikode). Single panchayats, municipalities and corporations cover areas of about 20 to 70 square kilometers and are further divided into *wards*. The panchayats are governed by elected presidents and executive members and administered by government employees. Apart from this political-administrative territorial division, a separate system exists for revenue collection. The 14 districts are divided into 61 *taluks* and over 1,400 *revenue villages*. Revenue village and panchayat borders often overlap.

Nearly all of the inhabitants of Kerala speak *Malayalam*, a Dravidian language with Sanskrit influences. Accordingly, the people of Kerala are also called *Malayalees*. The 1991 Census data breaks the population into:

- Hindus (60 percent);
- Christians (21 percent);
- Muslims (19 percent).

The proportion of non-Hindus, particularly of Christians, is relatively high compared to the national average. Still, communal harmony is further developed than in most other parts of the country. In the aftermath of the Ayodhya incidence, however, Hindu-Muslim clashes broke out in Kerala too, and 11 people were killed. As elsewhere in India, people responded with manifestations of "communal" (religious) tolerance. "Kerala's image of caste and religious tolerance was tarnished, but significant actions were underway to regain it" (Franke & Chasin 1994: 23). Also, the right-wing, "communalist" Hindu party BJP has not been able to get as strong a foothold in Kerala as in many other states of India.

Kerala is one of the least caste-ridden societies in India. This is remarkable because, in the early 20th century, Kerala still had the most rigid caste system, which included the concept of pollution by sight as well as untouchability. (The English term "caste" denotes two distinct caste categorizations: *varna* (literally, color) and *jati* (literally, birth). Varna refers to the broad hierarchical categorization into *Brahmins* (priests, landlords), *Kshatriyas* (warriors, administrators), *Vaishyas* (artisans, traders) and *Sudras* (cultivators, servants). Apart from this system stand the *avarnas* (non-caste Hindus including "Untouchables"). Each varna is divided into numerous hierarchical jatis, which often have local manifestations. The jati is the caste category people themselves refer to.)

Hinduism became the predominant religion in Kerala between the 8th and 11th century A.D., when Aryan Brahmins migrated from north India to areas dominated by Dravidians. Today, the most important castes (*jatis*) of Kerala are:

- *Namboodiri* (Brahmins: about 2 percent of Kerala's population);
- *Kammalan* (Vaisyas: 7 percent);
- *Nair* (Sudras: 16 percent);
- *Ezhava* (avarna: 22 percent);
- *Pulaya* and *Cheruma* (avarna, "Untouchables": 8 percent);

- *Adivasi* (tribal populations, not a caste in a strict sense: below 1 percent) (R.R. Nair 1976, cited in Franke & Chasin 1994: 91; figures are for 1968).

To describe a complex system very simply, Namboodiris traditionally were priests and landlords; Kammalans were artisans; Nairs were warriors, administrators and cultivators; Ezhavas were coconut-tree climbers; and Pulayas and Cherumas were dependent farm and menial workers. Since the early 20th century, the caste system has become more flexible in Kerala. Nowadays, ascribed occupational positions are not followed as strictly as in the past, and formal education has become an increasingly important factor in the choice of profession. Furthermore, low castes have gained in social and economic status. For example, the Ezhavas have risen at the expense of the Nairs.

Kerala is known for its absence of an indigenous Hindu trader caste. For centuries, the position of "traditional" traders has been taken by Muslims and Christians. Because of the early outward orientation, Christianity and Islam were introduced in Kerala long ago. Syrian Christians have lived in Kerala at least since the 5th century (some of their descendants claim that their origins even go back to a mission of Apostle St. Thomas in 52 A.D.), and Muslim traders and missionaries brought their religion to Kerala in the 7th century. In the course of time, Christianity and Islam have been included in, and have adapted to, the hierarchical Hindu social system (see Visvanathan 1986). Syrian Christians, for example, are like a middle caste and generally respected as good farmers. Latin Catholics or Protestants, on the other hand, are generally low-caste Hindus converted by colonial missionaries. Consequently, they are still treated as low-caste people and often belong to the poorest section of Kerala's society.

5.1.3 Climate, Natural Divisions and Environmental Condition

Kerala lies in the tropics between 8° and 13° northern latitude. Its climate is tropical, with fairly high temperatures and abundant rainfall. Seasonal temperatures do not vary much compared with daily variations. The average daily maximum temperature in the plains is around 31 °C, the average minimum temperature 24 °C. The coldest months are December and January; the hottest are March, April and May. Most parts of the state get two rainy seasons: the south-west monsoon from June to September and the weaker north-east monsoon in October and November. The dry season lasts about six months (from December to May). The annual rainfall, of more than 3,000 mm, is very high. About 90 percent of the rain falls during the monsoons. Generally, the northern districts have the longest and most pronounced dry season but very massive rainfall during June and July. In Central and South Kerala, the seasonal rainfall is more equally distributed and some rain may also fall in the summer months. Thanks to the high rainfall, groundwater resources are abundant. Furthermore, numerous perennial streams and 44 big rivers flow through Kerala, 41 of them toward the Arabian Sea (Kerala State Gazetteer 1986: 289-297).

Kerala's land between the Arabian Sea and the Western Ghats is usually divided according to altitude into three broad natural divisions that run north-south, namely, the lowland, the midland and the highland. According to the common definition in Kerala, lowland is below 7.6 meters above m.s.l., midland is between 7.6 and 76 meters above m.s.l. and highland is over 76 meters above m.s.l. In this classification, however, the category highland encompasses very different agro-ecological zones (including areas of an altitude of more than 2,000 meters). Therefore, the lower part of the highland (with an altitude of 76 to 400 meters above) is also called "mid highland." In this study, I include this lower part of the highland into the category midland because topography, climate, soils and cropping pattern are very similar.

The *lowland* lies on the western fringe of the state along the seashore and covers 10 percent of Kerala's total geographical area. It includes the 580-kilometer-long coastal belt with stretches of sand, lagoons (backwaters) and a huge area of reclaimed land (Kuttanad). Soils are generally rather poor here (sandy alluvial soils, various saline soil types). The *midland* accounts for 60 percent of Kerala's area. In most parts, the midland is an undulating terrain with rolling hills that are intercepted with valleys. In general, the laterites of the upland (dryland) as well as the hydromorphic soils of the valleys (wetland) are moderately fertile. In the midland, a large variety of perennial, annual and seasonal crops are grown. The *highland* lies in the east of the state and is part of the Western Ghats. It includes steep, mountainous terrain but also fairly even high plateaux. Here, forest soil is the most common soil type. The highland is covered with forests and plantations (Kerala State Gazetteer 1986: 39-44; Centre for Earth Science 1984).

According to comprehensive reports on the state of India's environment (World Resource Institute, 1994; Centre for Science and Environment, 1985), Kerala seems to suffer less than most other regions from environmental problems such as: deforestation; industrial pollution of air, water and soils; loss of soil nutrients, groundwater depletion and pollution due to intensified agriculture; salinization and waterlogging because of canal irrigation.

Although Kerala does not face a severe ecological crisis, environmental problems are apparent and affect environmental sustainability. The most important environmental problems are caused by former deforestation, ongoing paddy conversions and disruption of backwater ecosystems. Of growing concern are also the "chemicization" of agriculture, pollution of water and soils, "rurbanization" and air pollution by growing traffic. A range of other more localized events, such as industrial pollution in particular areas, excessive sand mining and pollution of some rivers, and destruction of natural *shola* grasslands with eucalyptus plantations, also trouble Kerala's generally good environmental record. Agriculture-related environmental changes and their implications for sustainable development are further discussed in Section 5.3.

5.2   Development in Kerala

5.2.1   Unique Development Pattern

Kerala has a unique development pattern: Despite its poverty in terms of economic indicators of development, the state displays a set of *very high social indicators* that are outstanding in comparison with the rest of India. This *development paradox*, or the corresponding redistributive policies, are referred to as the *Kerala model*.

Table 2 indicates that Kerala's social development (e.g., health, education) is very "advanced" in the Indian context. Furthermore, Kerala is also fairly advanced with respect to public services and infrastructure. Already in 1980, 91 percent of all villages had health dispensaries, more than 96 percent had drinking-water facilities and electricity, and more than 98 percent had primary and secondary schools, ration shops, all-weather roads, bus stops and post offices (Central Statistical Organisation, cited in Franke & Chasin 1994: 42). Kerala's life expectancy, infant-mortality rate and literacy rate do not lag much behind those of industrialized countries. At least in the case of Kerala, many Western stereotypes of the Third World are inaccurate (e.g., population explosion, inhuman lives and mass starvation, illiteracy, general powerlessness and hopelessness).

Table 2   Selected Development Indicators (Kerala, India)

	Kerala	India	Rank*
Demography:			
Population, 1991 (in million)[a]	29	846	
Decennial Population Growth Rate, 1981-90[a]	14	23.5	lowest
Total Fertility Rate, 1991 (children per female)[b]	1.8	3.6	lowest
Economic Development:			
Per Capita Net State/National Domestic Product (at current prices, 1991-92, in Rs.)[c]	4,618	5,583	10th
Growth Rate of Per Capita Net State/National Domestic Product, 1980-90 (percent per year)[d]	0.3	3.1	16th
Poverty and Unemployment:			
Rural Persons under Poverty Line (in percent)[e]	16	30	N. A.
Rate of Unemployment, 1987[f]	26	N. A.	17th (highest)
Social Development:			
Physical Quality of Life Index, 1991[g]	91	55	1st
Infant Mortality per 1,000 live births, 1994[b]	13	73	1st (lowest)
Life Expectancy at birth, 1992[b] Males	69	59	1st
Females	74	59	1st
Literacy Rate, 1991[h] Male	94	64	1st
(age 7 and above) Female	87	39	1st
Total	91	52	1st

*Rank of Kerala among 17 Indian states.

Sources: [a]*1991 Census, cited in Prakash 1994: 45-46*

[b]*Sample Registration System Data, Government of India, cited in Drèze & Sen 1995*
[c]*Economic Survey, Government of India, cited in Drèze & Sen 1995*
[d]*Central Statistical Organisation, Government of India, cited in Drèze & Sen 1995*
[e]*Official Estimates as in the Five-Year Plans, cited in Pillai 1994: 168*
[f]*Directorate of Economics and Statistics, Government of India, cited in Pillai 1994: 174*
[g]*Pillai 1994: 56*
[h]*1991 Census, cited in State Planning Board, Economic Review 1995: 131*

Already in the 1950s and 1960s, the indicators of social development were good in Kerala compared with other Indian states. Since then, much more progress has been achieved. Between 1961 and 1991:
- the Physical Quality of Life Index increased from 59 to 91 (Pillai 1994: 56);
- life expectancy increased from 48 to 72 years (State Planning Board, Economic Review 1995: 144);
- the literacy rate rose from 55 to 91 percent (State Planning Board, Economic Review 1995: 131);
- the infant-mortality rate fell from more than 60 to 13 (State Planning Board, Economic Review 1995: 143-144);
- the percentage of people living below the poverty line declined from 63 to 16 percent (official estimates, cited in Pillai 1994: 168).

Among these many achievements, the ones in education are particularly noteworthy. School enrolment is near to 100 percent and drop-outs are relatively few. In addition, many parents send their children to the mostly private English-language schools, and to additional tuition classes. Furthermore, many young people go for higher education. 160,000 Keralites study at university level, for example (State Planning Board, Economic Review 1995: 137). Generally, formal education has become an important cultural value in Keralan society. Good education for women also appears to have had a positive influence on children's health and reduced the rate of population growth. Furthermore, formal education has contributed to higher social mobility of low-caste people and to better opportunities for migration (Franke & Chasin 1994: 75). However, the comparatively good formal education system seems not to have given much impulse for Kerala's general economic development (see below).

Unlike social development, the economic performance of Kerala is not rosy. Table 2 shows that economic growth and per capita state domestic product (SDP) are far below the national average. In 1994-95, the per capita SDP was estimated to be about Rs. 8,200 (US$ 270) (State Planning Board, Economic Review 1995: 16). However, consumer expenditures are relatively high compared with other Indian states. This paradox is mainly attributable to the influx of *remittances*, which are very substantial for Kerala. In an adjusted estimate considering remittances from the Middle East, income from home gardens, etc., Van Kerm (1997) calculated the "real income" in Kerala to be about 45 percent higher than the per capita SDP. According to this adjusted calculation, Kerala would be ranked fifth among Indian states – much higher than according to the per capita SDP.

Since 1973, large numbers of Keralites have migrated to the Persian Gulf in search of work. In the 1980s, about 500,000 Keralites worked in the Gulf, most of them sending money back home (G.P.R. Nair 1994: 105). Most migrants come from poor households, for which Gulf remittances often form the main source of income. According to varying estimates by G.P.R. Nair & Pillai (1994), Oommen (1993: 156) and Gulati & Mody (cited in Mohandas 1994: 88), these remittances may be as much as 13 to 28 percent of Kerala's SDP, two to three times higher than the plan outlay in over 40 years (government expenditures for new development schemes as formulated in Five-Year Plans).

The influx of Gulf money may have contributed substantially to the decline in poverty. However, newly rich Gulf returnees have often used their money to catch up with consumption rather than to invest in production. Except for the construction sector, therefore, the inflow of Gulf money has not triggered economic growth. It may even have had negative consequences, such as increased imports from other Indian states resulting in huge trade deficits (Isaac 1994: 392), an increase in land prices, the spread of consumerism, a steep increase in dowry prices, and partial transformation of dowries (on which the married woman keeps entitlements) into "groom prices" (on which only the groom's family has entitlements).

5.2.2   A Model of Development?

Kerala's unique development pattern and its outstanding accomplishments, achieved without much foreign aid, have gained respect in international circles. Kerala's development has become known as the *Kerala model of development*, a designation originally put forward by the Centre for Development Studies:

> The fact that Kerala is a relatively poor state in India when judged by conventional norms such as per capita income; that the average per capita availability of food is lower in Kerala than in most parts of India; but that nevertheless it has been possible for the State to make fairly impressive advances in the spheres of health and education, and hence bring about improvements that have made a perceptible difference to the quality of life – as also the acquisition of attitudes and skills that could help to accelerate development at the next stage – has certainly some lessons for similar societies seeking social and economic advance. (Centre for Development Studies 1977: 153)

Kerala's success has been achieved through social and redistributive policies such as (see Centre for Development Studies 1977: 153):
- effective public distribution of food;
- systematic extension of public health and education facilities;
- effectual land reforms.

Popular movements generally struggled effectively for these welfare policies and radical reforms (Franke & Chasin 1994: 31-32). For the implementation of radical reforms, the first communist government of Kerala (1957-59), which, except for the tiny principality of San Marino, was the first ever democratically elected communist government beyond municipal level in the world, and successive Communist Party (Marxist)-led coalition governments (1967-69; 1980-82; 1987-1991; since 1996) have also played a crucial role. But Congress-led and -supported governments have also implemented social reforms, partly because they were pressured by leftist popular movements, peasant and labor associations.

However, some now question whether the Kerala model of development without economic growth really represents a *model*, and also whether social development in Kerala is the product of recent state policies and of political mobilization for radical reform. Kerala's achievements are not the direct outcome of a development *plan* and therefore can hardly be called a "model" in the sense of a design. While acknowledging the role of popular movements for Kerala's development, the proponents of the Kerala model generally neglect the influence of remittances. Moreover, significant progress had already been achieved during the 19th and early 20th centuries, particularly in the princely states of Travancore and Cochin. The princely states implemented land and tenurial reforms, built up health facilities and supported popular education (beyond the "traditional" philosophical and religious education system of Brahmins) in order to meet the requirements of commercializing export-oriented agriculture. The many European missionaries also established schools and hospitals. In addition, several caste-based reform movements took aim at the rigid caste system (Tharakan 1983 & 1992, both cited in Törnquist 1995: 20-21). Finally, Franke & Chasin (1994: 50) argued that the peculiar rurban settlement pattern may have helped to provide cost-effective health and educational services. While I agree that a dense population makes it easier to supply health and education facilities, I would argue that this would be still easier in a clustered rather than a rurban settlement pattern. On the other hand, the dispersed settlement pattern impedes the transmission of contagious diseases and epidemics.

Because of these particular historical and geographical conditions, Kerala cannot offer an easily reproducible model for other developing countries. Furthermore, die-hard protagonists of the Kerala-model idea have also overlooked the admission of the Centre for Development Studies (1977) that, for Kerala itself, the tasks ahead were still considerable (Mohandas 1994: 91). Since the 1980s, however, more attention has been paid to the failures of Kerala's development. Meaning to provoke, Prakash (1994) even described Kerala as a negative model:

> The basic weakness [of the Kerala model of development] is that it treated only a few demographic indicators as the indicators of the quality of life. But if we take unemployment of labour force and the extent and magnitude of poverty as indicators of quality of life, then the model becomes a *negative* model. (Prakash 1994: 28-29; emphasis not in original)

Moreover, the macro-indicators of social development hide regional and class disparities in development. Certain sections of the society, such as the fisherfolk, the Adivasis, elderly women and widows, have not been able to improve their situation and remain relatively oppressed. A micro-level survey of more than 10,000 households has revealed that there is also wide disparity in access to basic facilities such as toilets and safe drinking water (Iyer 1996). Generally, living conditions for a considerable percentage of the population are still deplorable. Kerala's very high and still rising suicide rate among young people also indicates that this society is far from devoid of social problems.

5.2.3 Economic Stagnation, Unemployment and Fiscal Crisis

Table 2 shows that Kerala's economic performance is feeble and that the economy is stagnating. Industrial development, for example, lags behind other Indian states. While India's manufacturing sector accounted for 20 percent of the Net Domestic Product (NDP) in 1994-95, Kerala's manufacturing sector accounted for only 16 percent and was also growing less rapidly (State Planning Board, Economic Review 1995: S1-S3). In the 1970s and 1980s, furthermore, Kerala's traditional industries (coir, i.e., the fibre of coconut husk, cashew and handloom textiles) shifted to other states, and new industries were not able to create sufficient employment. Generally, Kerala's industrial backwardness is explained with comparatively high production costs due to high input and labor costs, lack of adequate electrical power, and unhealthy practices of trade unions (Prakash 1994:

20). On the other hand, the construction sector, transportation and trade are growing at a relatively fast pace.

Another feature of Kerala's economic failure is agricultural stagnation (see Section 5.3). Marked since the mid-1970s, agricultural stagnation can be traced back to such factors as inefficient use of irrigation, insufficiency of inputs, and lack of long-term investment in land improvement (Sivanandan 1994: 156), implementation of ill-suited large irrigation schemes, previously almost unknown summer droughts, neglect of local resource management, absence of regulated markets, and failure to promote cooperation among farmers (Törnquist 1995: 43, 59-60). Agricultural production and productivity showed an upward trend from 1991 to 1994, but this was probably a consequence of favorable weather conditions (State Planning Board, Economic Review 1995: 5). With the failure of the north-east monsoon in 1995 and of the south-west monsoon in 1996, agricultural production may have received a setback again.

The general economic stagnation is explained in very different ways:

> [According to the rightist theorists] ... the trade unions and high wages caused the retarded industrial growth. The faulty over-growth of education was responsible for the problem of educated unemployment. The welfare measures were responsible for the fiscal crisis of the state. Severe restrictions of the social welfare expenditure and privatisation of public services was put forward as the only alternative to the resource crunch of the State Government.
> The leftist response was mostly to emphasise factors such as the historical roots to industrial backwardness of Kerala, the uneven development of crisis-ridden capitalism in India, distortions in Centre-State relations, discriminatory policies of the centre towards Kerala, particularly in the allocation of institutional finance and central public sector investment, the central agricultural price policy and so on.... (Prakash 1994: 28)

Remarkable is also the contrary assessment of the role of Kerala's high degree of "politization." Prakash (1994: 36-38), for example, argued that unfavorable political and labor factors (e.g., militant trade unions, numerous strikes) have discouraged investment in the primary and secondary sectors. Franke & Chasin (1994: 32), on the other hand, insisted that the social reforms would not have been possible without militant and organized struggle. In my opinion, people's awareness of their rights and popular actions to improve concrete living conditions have been crucial for Kerala's social development. However, it is equally certain that "party-politization" and agitations undertaken for the sake of opposing the ruling party have resulted in high economic loss and contributed to stagnation in Kerala. Also, the cyclical pattern of politics in Kerala (i.e., the successive changes of government) has hindered the formulation of continuous and consistent policies whose implementation might have reduced uncertainty among Kerala's entrepreneurs.

Economic stagnation has had severe consequences for the employment situation in Kerala. Although data on unemployment from different sources differ widely and may not be very reliable, it is certain that unemployment has increased significantly since the formation of Kerala: according to the Directorate of Economics and Statistics (cited in Pillai 1994: 174), the rate of unemployment rose from 9 percent in 1965 to 26 percent in 1987. By 1995, 3.2 million Keralites were looking for a job (State Planning Board, Economic Review 1995: 17). Unemployment is particularly high among formally educated people (persons older than 15 with secondary-school qualification or more). This widespread phenomenon is referred to as *educated unemployment*. Generally, the curricula of formal education do not match the needs or potentials of Kerala's economy. Higher education, for example, prepares the students only for government or teaching jobs. Consequently, the highly educated unemployed seek government jobs, which are also associated with social security and status. They become job seekers rather than job creators (Oommen 1993: 118-125; Prakash 1994: 35-36). Moreover, the preference of young formally educated people for white-collar jobs partly explains the paradox of labor scarcity in agriculture and the high incidence of rural unemployment.

Other consequences of Kerala's economic stagnation are the increasing scarcity of financial resources to pay for costly welfare schemes such as pensions, unemployment relief and the public (food) distribution system (K.K. George 1993). The fiscal crisis together with the underdevelopment of productive sectors and the high reliance on Gulf money increasingly set limits to, and threaten the *sustainability* of, the so-called Kerala model with its redistributive policies and reforms. Generally, Kerala's development priorities of the 1990s are to strengthen the production basis and to achieve economic growth in order to overcome unemployment and remaining poverty. For that, Oommen (1993: 211), for example, proposed an export-led growth strategy, enhanced linkage between agriculture and industry, infrastructural improvements (particularly in the power sector), and other such reforms.

In accelerating growth and strengthening the production basis, India's *New Economic Policy* of 1991 may become important. However, economic liberalization and deregulation under the New Economic Policy have not been fully implemented yet and tend to be only halfhearted. Particularly in the agricultural sector, many government regulations such as price controls, state procurement and non-tariff trade restrictions are still practiced (see Randhawa 1994: 356-357). Also, fertilizer and food subsidies were reintroduced before the 1996 national elections. Because of "incomplete" implementation and the irregular path of economic reforms in India, their consequences for Kerala's (agricultural) development and environment are difficult to assess. At the same time as the economy is becoming globalized, Kerala's development planning is being decentralized. The *Panchayati Raj Act*, which was revised in 1994 in order to conform to amendments in India's constitution, involves the transfer of power, responsibilities and funds from state departments to the panchayats in order to initiate decentralized, local development planning with the support of non-governmental organizations and people's participation.

5.2.4 Sustainable Development?

Despite the above-mentioned economic problems that may threaten the sustainability of Kerala's welfare policies, some (mostly Western) scholars have transcended the standard arguments for the Kerala model to call Kerala also a model for environmentally sound, sustainable development. Obviously, Kerala fulfils some core objectives of the concept of sustainable development, such as low population growth rates and moderate industrial pollution. Kerala's per capita energy consumption and use of natural resources are still very low compared with "developed" countries. Therefore, Alexander (1994) regarded Kerala as an example of "efficient resource use," "modest consumption" and "sustainable human behavior." McKibben (1995) pointed in the same direction:

> Kerala suggests a way out of two problems simultaneously: not only the classic development goal of more calories in bellies and more shoes on feet, but also the emerging equally essential task of living *lightly* on the earth, using fewer resources, creating less waste. (McKibben 1995: 163)

However, Alexander's and McKibben's analyses neglect the spreading consumerism, the influx of Gulf money, the growth in imports of consumer goods and resources. They also fail to take the mentioned "actually existing" environmental problems in Kerala into account. Furthermore, Kerala's generally good environmental record originates from natural peculiarities to some extent: the hilly topography has hindered the wide expansion of environmentally unsound Green Revolution technologies; comparatively high and seasonally well-distributed rains contribute to low sensitivity and high resilience of the land. Also, comparatively low levels of industrial pollution is a consequence of industrial backwardness rather than of *clean* industries. Moreover, "living lightly on the earth," "modest consumption" and "sustainable human behavior" in Kerala unfortunately relates more to poverty than to sustainable *development*:

> Like anywhere else, the rich in Kerala live "heavily" on the earth.... Those people who live lightly on the earth, "sleeping on the floor", without many possessions, do so, not out of ecological consciousness, as one *[particularly Western environmentalists]* would like to believe, but because they cannot afford even the essential possessions. (Iyer 1996: 216)

Economic underdevelopment and bio-physical factors, rather than a particular model of development, have influenced the relative environmental sustainability. Environmental concerns have only played a secondary role in Kerala's policy-making (which, however, may change with decentralized development planning). Furthermore, economic stagnation and fiscal crisis are threatening the continuation of the welfare policies that have been important for Kerala's remarkable social development. Although, at first sight, Kerala may come reasonably close to, and show some features of, the sustainable-development ideal (Parayil 1996), it cannot be called a model for sustainable development.

5.3   Agriculture and Sustainability

5.3.1   Economic Significance of Agriculture

Despite its declining share, agriculture still accounted for 29 percent or Rs. 62 billion (US\$ 2 billion) of Kerala's SDP in 1994-95 (State Planning Board, Economic Review 1995: S3). Furthermore, the agricultural sector is closely linked to Kerala's industry. Agro-processing industries (e.g., food processing, cashew, coir, rubber, tobacco) are very important, particularly in terms of employment; and a substantial proportion of trade and commerce is in agricultural commodities. Spices, cashew, tea and coffee, marine products and coir are also the most important export items of Kerala, accounting for 85 to 90 percent of the state's foreign exports (Isaac 1994: 371).

Agriculture remains a major source of employment. In the 1991 Census, about 1,100,000 Keralites (12 percent of the workforce) were counted as cultivators, and more than 2,400,000 persons (27 percent of the workforce) were counted as agricultural laborers (cited in Pillai 1994: 41). Furthermore, agriculture represents an important source of subsidiary or seasonal income for many households in Kerala. Construction workers, quarry workers, petty traders and even teachers, government employees and politicians are engaged in agricultural production or retain agricultural land.

5.3.2   Cropping Pattern and Agricultural Change

According to the Directorate for Economics and Statistics (Statistics for Planning 1995), Kerala's land use breaks down into:
- "net area sown" or land under cultivation (58 percent of the geographical area);
- "forest land" (official category that also includes non-wooded land under the management of Kerala's Department of Forestry) (28 percent);
- land under non-agricultural use (8 percent);
- fallow and waste land (4.5 percent);
- uncultivable land (1.5 percent).

The "total cropped area" (net area sown plus the area that is cultivated for a second and third time in the same year with seasonal crops) is about 7,630,000 acres (one acre = 0.4046 hectares). According to Kerala's Directorate for Economics and Statistics (Statistics for Planning 1995), the most important crops grown are:
- coconut (28 percent of total cropped area);
- paddy (18 percent);
- rubber (15 percent);
- fruit (8 percent);
- pepper (6 percent);
- cashew (4 percent);
- tapioca (cassava) (4 percent);
- other crops (17 percent).

A general weakness of agricultural statistics is that they break down by administrative units (districts) as opposed to natural divisions. However, cropping patterns show distinct features in the three natural divisions of Kerala (lowland, midland, highland). Of course, the division into lowland, midland and highland can also only provide a very simplified picture that neglects diversity within each of these natural regions. Kerala's Committee on Agro-Climatic Zones and Cropping Patterns (1974), for example, distinguished between 13 agro-climatic zones, each with its relatively distinctive cropping pattern.

Kerala's midland is the most important zone in terms of both area and population. In the midland, a huge diversity of crops are grown on the laterites of the upland (also called gardenland). Perennial crops include coconut, rubber, cashew, arecanut, cocoa, nutmeg, clove, jackfruit, mango, pineapple, pepper; annual crops include tapioca, banana (including plantain), ginger, turmeric; seasonal dryland crops include vegetables, tubers and pulses. Most of the dryland crops are grown in intensive mixed-cropping systems – in the numerous small home gardens and also on extended, mostly coconut-based, mixed gardens that lie further away from the house. Home gardens and extended mixed gardens are the major agricultural production systems in Kerala, accounting for more than half of the cultivated area (Jose 1991: 45-46). Almost every household has a home garden, where subsistence crops and cash crops are grown, livestock is reared, and fodder, fuel and timber are produced. In addition to their economic value, Keralites also attribute social, aesthetic and cultural meaning to the home gardens (Jose 1991: 2). However, some of the typical midland crops (e.g., coconut, cashew, tapioca, banana, pineapple, ginger) are not only grown in mixed gardens but also as pure crops. Furthermore, rubber, which is very common in Kerala's midland, is always grown in monoculture.

In the wetland valleys of Kerala's midland region, paddy fields are widespread. The hydromorphic soil and waterlogged conditions are most suitable for paddy cultivation. In well-irrigated parts, three paddy crops per year are possible; in other parts, vegetable or pulses are grown as the third crop, or the land is left fallow for three or four months. Since the mid-1970s, however, many paddies have been converted into dryland. For a long time, Kerala's irrigation schemes have focused on large-scale canal irrigation and on the irrigation of paddy in these low-lying areas. The government irrigation projects, however, were not implemented efficiently, do not run effectively and are poorly maintained. At present, greater potential is seen for minor irrigation schemes and for the irrigation of dryland crops. Apart from the government, private parties have initiated small irrigation schemes (see Jacob 1996).

In the lowland, the temporarily waterlogged land is normally used for cultivation of salinity-resistant paddy. During the monsoon months, when the paddies are inundated, many fields are also used for prawn farming. Between the paddies, coconut and arecanut groves are most widespread. For the people living in the lowland, maritime fishery (with traditional boats and motorized trawlers) and the use of backwater resources (fish, shrimps, clams, shells, retting ground for coconut husk) are also important (see Paul 1996). The major cities and towns of Kerala are located in the lowland, where as early as in the 19th century, an extensive water-canal system connected the various settlements from north to south. In the highland of the Western Ghats, by contrast, plantations of pepper, cardamom, tea, coffee and teak are most widespread, alongside remaining rainforest, mixed gardens and a few paddies.

Until the mid-1970s, Kerala's agriculture was continuously expanded and intensified. From 1960-61 to 1975-76, the net area sown expanded by more than 750,000 acres (8 percent of Kerala's total area) and the total cropped area grew by almost 17,500,000 acres. During that period, both agricultural production and productivity also increased. The agricultural extensification was mainly at the cost of forests in the highland, of land under miscellaneous trees (bamboo, casurina, groves for fuel, thatching grass), of permanent pastures and grazing land and of different types of fallow and waste land.

In the mid-1970s, however, agricultural growth in Kerala slowed down and, since then, it has even been negative in some years. Generally, the season of 1975-76 is considered the major turning point of Kerala's agricultural performance (Kannan & Pusphangadan 1988). This turning point coincides with a change in the methodology of data collection by Kerala's Directorate of Economics and Statistics (see G.S. Nair 1983). The change in methodology, however, cannot explain the downward trend of Kerala's agricultural performance since 1975-76.

The mid-1970s also mark the beginning of an accelerating shift from seasonal and annual food crops (i.e., paddy and tapioca) to perennial tree crops (i.e., rubber and coconut) and various export-oriented crops (i.e., pepper, ginger and coffee). Since the mid-1980s, the cultivation of fruit crops (banana, mango, jackfruit, pineapple, papaya) has also been increasing. Between 1975-76 and 1994-95, the area under paddy cultivation declined by 43 percent, and tapioca cultivation by 59 percent. Meanwhile, the area under rubber cultivation increased by 114 percent, coconut cultivation by 30 percent, pepper cultivation 75 percent, coffee cultivation by 95 percent and fruit cultivation by 29 percent. In addition to experiencing changes within agriculture, Kerala has also seen its agricultural production undermined by rurbanization. Between 1960-61 and 1993-94, the land put to non-agricultural use increased from 500,000 acres to more than 750,000 acres (P.S. George 1996: 4-7, based on data from Kerala's Directorate of Economics and Statistics).

In the highland of the Western Ghats, deforestation of natural rainforest has been the most significant land-use change. Deforestation was probably the most severe environmental change in Kerala until the mid-1980s. Based on satellite images and topographical sheets of the Survey of India, Chattopadhyay (1984) has concluded that Kerala's forest area declined from 44 percent of the total geographical area in 1905 to 28 percent in 1965, 17 percent in 1973, and 7 or 10 percent in 1983. Since then, deforestation seems to have stopped (Parayil 1996). However, the deforested highland continues to affect the environment as well as human well-being (see Section 5.3.3). Deforestation can be attributed to expanding plantations, government hydroelectric projects and agricultural colonization schemes, encroachments of land-seeking farmers, and illegal felling by timber contractors. Generally, deforestation is not primarily caused by population pressure but, among other factors, by particular state policies (Kannan & Pusphangadan 1988: 125). In the highland, changes have also taken place within the plantation sector. Govindan (1996), in a village-level study, showed that small plantation owners are very responsive to changes in international prices of plantation crops. In recent years, for instance, the major shift was from cardamom to pepper. This shift has also contributed to deforestation because trees must be felled for pepper cultivation whereas cardamom thrives as forest undergrowth (K.N. Nair *et al.* 1989).

In the midland, the most significant land-use change is the conversion of paddies. Since the mid-1970s, about 40 percent of the wetland paddies have been converted into dryland. Only the plain of Palghat district (Kerala's "rice bowl") has been spared from extensive paddy conversions. Usually, a temporary wetland conversion into annual crops such as banana or tapioca is followed by a permanent conversion into coconut and other perennial trees, or by the use of former paddies as dwelling sites. Furthermore, paddies are mined for clay, which is used for brick production (see Pronk 1997). The most significant land-use change on the upland of Kerala's midland region is the spread of rubber plantations, particularly in the southern part of the state.

In the lowland, backwaters have been widely reclaimed for residential, industrial and agricultural purposes.

5.3.3 *Environmental and Socioeconomic Impact of Agricultural Change*

In Kerala, agriculture accounts for 29 percent of the state domestic product and for 39 percent of the workforce. Consequently, agricultural stagnation has a major impact on the level of economic growth and employment. The significance of this slowdown in the agricultural sector is magnified by the slowness of industrial development and the failure of the growing tertiary sector to employ more than a small portion of the job-seekers. Strengthening Kerala's production base, a prerequisite for sustainable development, thus also depends on agricultural growth. On the other hand, agricultural growth, expansion and intensification can also contribute to resource exploitation and environmental pollution, thereby making development less sustainable. Industrial pollution in Kerala is comparatively low; much of Kerala's (fairly moderate) environmental degradation is caused by agriculture. Although agriculture-induced environmental degradation is not always evident on first sight, it increasingly affects people's livelihood in Kerala.

Changes in land use and agricultural practice have different socio-economic and environmental impacts. Deforestation in Kerala's highland, for instance, has led to massive loss of biodiversity and genetic resources, more landslides, enhanced soil erosion, faster water run-off, laterization and changes in the local climate (Kannan & Pusphangadan 1988: A125). As regards socioeconomic consequences, deforestation has directly affected the Adivasis, who formerly gained a significant part of their living from gathering forest products. Deforestation in the highland also has consequences for people in the midland and lowland. Many household wells in the midland, for example, are drying up and need to be dug deeper and deeper. Not only declining groundwater availability, but also the increased incidence of landslides and floods (that in some cases lead to evacuation and destruction of houses) can partly be ascribed to deforestation and accelerated water run-off. Furthermore, climatic change becomes manifest in increased incidence of droughts, which did not occur in Kerala before the 1980s. However, this climatic change may not be directly related to the deforestation in the Western Ghats but rather to overall changes in the regional monsoon system. The drying-up of wells is also related to paddy conversions and the spread of rubber plantations (see below).

The spread of rubber cultivation in Kerala's midland also has environmental and socioeconomic impacts. The partial replacement of the environmentally sustainable mixed gardens with rubber monocultures signifies a loss of biodiversity (Jose 1991: 122). The shift from pure tapioca cultivation to terraced rubber plantations on slopes, on the other hand, may have reduced soil erosion. Other environmental effects of rubber plantations are ambiguous. Allegations by local people that extensive rubber plantations resulted in increased local temperatures (caused by the high sunlight reflecting off the leaves) or depleted groundwater resources (because rubber is a water-exhausting crop that taps deep-lying groundwater resources) have not been verified using scientific methods (see also Narayanan 1996: 12-13).

The environmental consequences of the widespread paddy conversions are more obvious and severe. For the required drainage, canals are dug in the fields (temporary conversion); or the fields are filled with soil from elsewhere and the water is drained by means of deep canals and walls that are constructed upstream (permanent conversion). The enhanced drainage lowers the water table because water, which under waterlogging conditions would leak down to the groundwater table, is not held back on the fields. (Under paddy cultivation, the soil remains sufficiently permeable to leak down water, although the paddy fields are levelled before sowing and made less water permeable in order to create waterlogging conditions, which restrict weed growth.) Paddy conversions thus change the physical and chemical soil properties. Furthermore, water is channelled directly into the side-canal (the water level of which is below the level of the fields) and does not overflow to the next field (see Fig. 6). Eventually, paddies adjacent to higher-lying converted fields get less water. By contrast, because more water flows through the side-canal, its banks and bed become prone to erosion. In general, paddy conversions mean abandoning the sophisticated traditional wetland agrosystem. As a result, ecologically valuable wetland fauna and flora disappear and the biological food chain is disrupted.

Apart from their adverse effects on the environment, paddy conversions produce changed conditions for agriculture and people downstream. Conversion of higher-lying fields can render the continuation of paddy cultivation impossible because water becomes too scarce. Consequently, cultivators lower down are compelled to convert their paddy fields into dryland as well. Especially in the case of permanent conversion into dryland, the cultivators' future flexibility is obstructed: Reverting immediately to wetland-paddy cultivation becomes impossible. In the case of temporary conversion, it requires high investment. Moreover, the change of the wetland system results in lower water-retention on the fields; and more surface run-off increases the probability of floods downstream. In fact, even under normal rainfall conditions, the cities in Kerala's lowland experience severe floods more often than in earlier times. As another consequence of floods, large fields become waterlogged more often and yields decline.

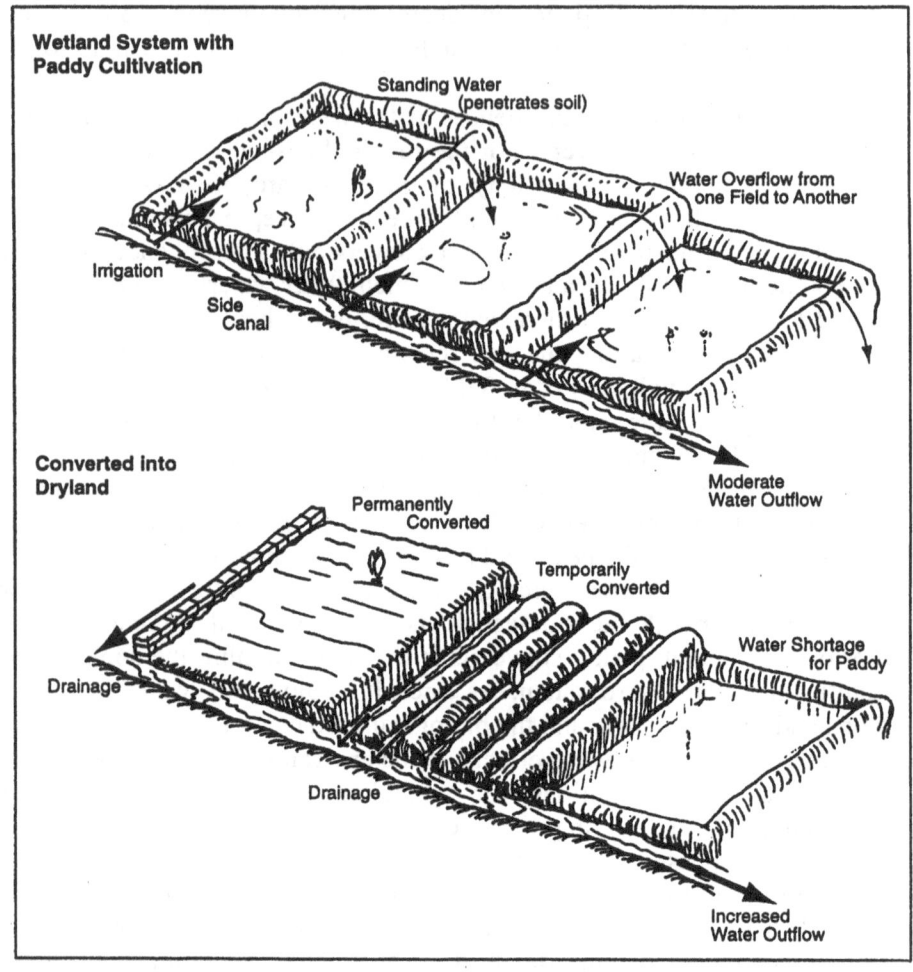

Fig. 6 Environmental Impact of Paddy Conversions
Illustration: E. Schmitt based on draft by R. Véron.

Decreased paddy cultivation – together with increased cultivation of coconut and rubber – have also worsened the employment situation for agricultural workers. Between 1963-64 and 1983-84, for example, available days
of employment for agricultural workers declined by 26 percent for males
and by 30 percent for females (Directorate of Economics and Statistics
1985, cited in Herring 1989: 107). The general shift from food crops to cash
crops has also made households more dependent on the market. Some

politicians and scholars in Kerala have argued that this may have severe implications for household food security as well as for Kerala's food security in general. Until now, however, it has always been possible to import rice from other states of India where paddy production is cheaper and more popular among farmers. Indeed, food consumption, even among poor households, has improved, thanks to increasing purchasing power and the distribution of rice at subsidized rates.

The increased use of agro-chemicals in Kerala has had negative consequences. Yet, the aggregate fertilizer use is far below the recommended dosage (M. George 1991). However, in certain regions and for particular crops such as rubber and paddy, too much chemical fertilizers and pesticides are applied. Moreover, fertilizer types and application times are often inappropriate. In some parts of Kerala, excessive use of agro-chemicals has destroyed the microbiology of the soil, and adversely affected soil recoverability and fertility. Moreover, excessive use of agro-chemicals has also polluted rivers, backwaters and drinking water (Unnikrishnan 1993; Verghese 1986). The destruction of mangroves in coastal areas as well as the reclamation and pollution of backwaters have not only hampered a unique ecosystem but also affected the livelihood of the fisherfolk.

The discussion above shows that changes in land use, cropping pattern and cultivation intensity have particular environmental and socioeconomic consequences. In turn, these consequences, in conjunction with other environmental and socioeconomic conditions, influence the land use and agricultural practice of farmers and, finally, Kerala's agricultural performance. Kannan & Pusphangadan (1988), for example, suggested that environmental degradation has been a main factor responsible for the agricultural stagnation in Kerala since the mid-1970s. Deforestation in highlands (and consequent soil erosion), ill-designed water-control projects in Kuttanad (and consequent growth of aquatic weeds, infertility of soils and waterlogging of paddy fields) and climatic change (rainfall deficiency, higher rainfall variability and reduction of rainy days) may have contributed to the dismal performance of agriculture (Kannan & Pusphangadan 1988: A125-A126). However, bio-physical conditions still allow cultivators many options regarding land use, crop selection and cultivation methods. My own field studies suggest that environmental changes have had less effect on the cultivators' practices and on Kerala's agriculture than have socioeconomic changes. The following section, therefore, focuses on agrarian relations that influence the cultivators' practices, agricultural performance and changes in cropping pattern.

5.4   Recent Changes in Agrarian Relations

5.4.1   Land Fragmentation and Land Reform

In Kerala, only 0.2 acres of cultivable land are available per capita. Table 3 shows that operational land holdings are very small. Average holding size within the smallest category (marginal holdings), which account for 92 percent of all operational holdings and 46 percent of the cultivated area, is only 0.45 acres.

Table 3   Operational Landholdings in Kerala, 1985-86

	Number of Holdings (Percentage)	Area Operated in Acres (Percentage)
Marginal Landholdings (below 2.5 acres)	4,473,000 (92%)	2,020,000 (46%)
Small Landholdings (2.5 to 5 acres)	281,000 (6%)	945,000 (22%)
Semi-Medium Landholdings (5 to 10 acres)	104,000 (2%)	670,000 (15%)
Medium Landholdings (10 to 25 acres)	25,000 (<1%)	325,000 (7%)
Large Landholdings (above 25 acres)	4,000 (<0.1%)	425,000 (10%)
Total	*4,887,000 (100%)*	*4,385,000 (100%)*

Source: Agricultural Census, cited in Farm Information Bureau, Farm Guide 1995: 23-24.

Most households with marginal holdings also have to engage in economic activities other than agriculture on their own land. My own field studies suggest that a small family needs a minimum of two to three acres to sustain a modest livelihood completely from cultivation on land it owns. In Kerala, "small" landholdings (2.5 to 5 acres, according to the all-India definition) are already considered fairly big, and "semi-medium" land-holdings (5 to 10 acres) or "medium" landholdings (10 to 25 acres) are considered large. "Large" holdings (more than 25 hectares) in Kerala are operated by a very few corporate plantations and a few big private plantation owners.

The small holdings of Kerala become further fragmented through population growth and through partitions by inheritance. In a village case study in south Kerala, it was found that, on average, every operational holding had been divided into 3.9 parcels during the preceding 20 years (Jose 1991: 84). Usually, land is passed on to sons and daughters equally. Compensation in cash is rather the exception. Land may also be transferred in the form of a dowry. Changes in inheritance institutions have also contributed to the fragmentation of holdings. The matrilineal system of inheritance *(marumakkathayam)*, and the joint-family system among the Nairs and other important landholding castes, gradually vanished during the first half of the 20th century. Formerly, land of the Nairs was not partitioned but remained undivided property of the *tharavadu* (matrilineal clan household). Today, the Nairs practice the nuclear-family system and land partitions by inheritance (Saradamoni 1983: 38). The other important traditional landed caste, the Namboodiris, maintain their customary inheritance system that passes the entire land on to the eldest son. In the course of the land reform of 1970, however, the former Namboodiri landlords lost most of their land.

Kerala's government has imposed several regulations on the ownership of land, land transfers and cropping pattern. Apart from the influential land reform of 1970 (see below), Kerala's Land Utilisation Act is worth mentioning. The Land Utilisation Act of 1967 prohibits the utilization of paddy land for other purposes than paddy cultivation without the permission of the District Collector. However, this order has not been seriously implemented, as the high incidence of paddy conversions indicates (P.S. George 1995: 3-5).

The *land reform* of 1970 is probably the most influential recent change in regard to agriculture in Kerala. Kerala's land reform is one of the most thorough, radical and best-implemented in South Asia. Prior to the land reform, Kerala had the third most unequal distribution of land in India and was characterized by a complex multi-tier tenure system that was interlinked with the caste structure. To put it simply, the tenure system consisted of (see Franke & Chasin 1994: 76-83):

- big landlords *(jenmis,* mainly from Brahmin castes);
- superior tenants *(kamandars,* high-caste Hindus, Syrian Christians and upper-class Muslims);
- cultivating tenants *(verumpattamdars,* "tenants at will" of the superior);
- agricultural workers (Pulayas and other "Untouchables").

The tenure system showed high regional variation: It was least pronounced in Travancore and most pronounced in the paddy regions of South Malabar. In 1957-58, the agricultural population in Travancore was divided into 56 percent owner-cultivators, 25 percent tenant-cultivators, and 13 percent landless laborers. In Cochin, 29 percent were owner-cultivators, 50 percent tenant-cultivators, and 19 percent landless laborers. In Malabar, only 10 percent were owner-cultivators, 75 percent tenant-cultivators, and 12 percent landless laborers (Varghese 1970: 161). A colonial official called Malabar "the most rack-rented place on the earth." In addition to paying high rents (mainly for wetland paddies), the cultivating tenants also faced insecurity of tenure. Moreover, both tenants and laborers had to fear eviction from their hutment dwellings on the landlords' holdings *(kudi-kidappu)*. The threat of eviction enhanced the power of landlords over tenants and laborers considerably (Herring 1983: 153-179; Logan 1882, cited in Kurup 1981: 24-34).

For decades, peasant movements and political parties (particularly the communists) struggled against these exploitative feudal agrarian relations. In 1970, the Kerala Land Reform (Amendment) Act was finally implemented; thanks to a committed state government and constant pressure from below, the land reform was nearly "completed" in the 1980s. The most fundamental component of the 1970 Act was to abolish landlordism and tenancy. Since 1970, about 37 percent of the net area sown has been transferred to about 1.3 million former tenants (43 percent of rural households). Most of these are smallholders (89 percent with holdings below 2.5 acres; 67 percent with holdings below 1.25 acres). However, relatively big cultivators with more than five acres of land, who formerly had belonged to the privileged class of *kamandars*, have received about 64 percent of the transferred area (Oommen 1993: 5; Franke & Chasin 1994: 79- 83; Herring 1983: 180-216).

Apart from the abolition of landlordism, the land reform aimed to redistribute land to landless people. Therefore, a maximum land ceiling of 20 acres for large families (maximally six to seven acres per person) has been introduced, although plantations of rubber, tea, coffee and cardamom are exempt. However, big landowners bypassed land ceilings through bogus land transfers within the family. Moreover, the exemption for plantation crops has reduced the availability of redistributable land. Most of the big landlords had leased out their land in former times, and the land reform simply allotted the land to their existing tenants. Consequently, the reform in this case, too, did not add to the supply of land available for redistribution. For one reason or the other, only about 175,000 acres of surplus land has become available for redistribution to landless households. Furthermore, this land was often of poor quality, and so was often abandoned or sold by the households to which it was allotted (Oommen 1993: 9).

In another important provision of the Land Reform Act, landless laborers whose huts were on the landlord's land *(kudikidappukars)* were granted heritable security of tenure and the right to purchase house-compound land at a very low price. In this way, about 300,000 landless households have become owners of their house-compound land. All in all, only 8 percent of rural households in Kerala have remained landless (Oommen 1993: 5; Franke & Chasin 1994: 78-79).

Despite the successful abolition of rents and of a parasitic class of landlords, Kerala's land reform did not have the expected positive impact on agricultural production and productivity. On the contrary, agricultural growth has declined since the implementation of the 1970 land reform. A major reason for this may be that most of the land was not transferred to small working farmers. Rather, many of the new landowners have never cultivated their own land but have been *supervising* tenants who employed laborers and sub-leased land. The Kerala Land Reform Survey of 1968, for example, revealed that on small landholdings (2.5 to 5 acres), 76 percent of the labor was hired and only 24 percent of the agricultural work was carried out by family labor. Even on micro-holdings below one acre, 47 percent of the labor was hired. So, despite the slogan of the land reform ("land to the tiller"), a major portion of the land has not been passed to the actual tillers of the soil, the agricultural laborers (Herring 1983: 180-216; Oommen 1993: 4-5; Törnquist 1995: 28).

Even more important is the fact that the majority of the new owner-cultivators have no clear-cut interest in agriculture. Most of them are part-time cultivators with other significant sources of income besides farming. These part-time cultivators tend to avoid high investments in intensive agriculture – middle-class and rich cultivators (e.g., employees, teachers) because they are not much interested in optimizing agricultural production; poor peasants (e.g., wage laborers) because they lack the means to invest. Statistics indicate that part-time farming has become more widespread in Kerala since the implementation of the land reform: The proportion of people whose primary income comes from cultivation fell from 21 percent of the total workforce in 1961 to 14 percent in 1981 (Census of India, cited in M. Eapen 1994: 68).

The growth of part-time farming has two main causes. First, many people shifted to part-time farming because the land reform and land fragmentation by inheritance produced very tiny holdings that are not sufficient for a household to meet all its needs from agriculture on land it owns. Second, many people have shifted away from agriculture as the main economic activity because of better business and income opportunities in the tertiary sector (e.g., trade, government jobs). Farming has become less of a primary business and self-employment opportunity and more of a reliable secondary source of income. Still, these people retain their farming interest and hire labor to work on their land. Consequently, part-time farming (as opposed to full-time farming) is not only practiced on small homesteads, but also on many fairly large holdings. Parallel to the declining economic importance of agriculture, however, has been a lessening cultural attachment to agriculture.

It should also be noted that for many middle-class and rich landowners, land has become a valuable asset and exchange commodity rather than a simple means of production. This trend of the *commoditization* of land has been enhanced by Gulf returnees who buy land for speculation. The influx of Gulf money into Kerala's land market has raised land prices very much – often beyond the reach of ordinary small farmers, who are thereby priced out of the land market. Between 1980-81 and 1984-85, for example, prices for paddy land increased more than 150 percent and prices for land under coconut cultivation increased more than 200 percent (calculations based on Oommen 1993: 22). As a consequence of the very high prices, land is not allocated automatically in the land market to those who want to, or need to, engage in agriculture as the main occupation. Agricultural-census data reflects that the concentrating effect of the land market has worked against the redistributive effects of the land reform. The area under large holdings (above 25 acres), for instance, increased from 7 percent in 1980-81 to 10 percent in 1986-87 (Oommen 1993: 7-8).

Landowners who regard land as an asset tend to cultivate less intensively and let land remain fallow. Moreover, the ban on tenancy hinders the leasing of land to people who have a genuine interest in agricultural production. Nevertheless, land is once more being leased in Kerala. However, it is generally leased only for short terms of less than one year, because landowners fear that lessees might otherwise claim title to the land by citing the illegality of tenancies and appealing to the "land-to-the-tiller" ideology. Such short-term leases are clearly not capable of generating a long-term interest in land productivity; lessees therefore tend to engage in short-sighted and unsustainable agricultural practices.

Land fragmentation, part-time farming and the commoditization of land have contributed to overall agricultural stagnation and adversely affected economic development. Furthermore, these peculiar land relations influence the cultivators' crop selection. Rich and middle-class part-time farmers tend to shift to perennial crops such as rubber and coconut that require less labor-time and are easier to supervise than seasonal crops. Also, the small-holding structure in wetlands discourages cultivators from continuing paddy cultivation, which requires coordinated water and land management practices in entire micro-watersheds.

5.4.2 Changed Labor Relations and Agricultural Wages

In Kerala, hired agricultural workers traditionally carry out the bulk of the work in the fields. Work participation of landowners is often limited to supervising wage laborers. Most landowners avoid manual labor, which they associate with social inferiority. In addition, the labor division is also pronounced among agricultural workers themselves, and goes beyond gender differentiation to caste-based differentiation. Unlike male and female Pulayas, for example, coconut-tree climbers (mostly male Ezhavas) or rubber tappers (mostly males) rarely engage in paddy or general agricultural work such as clearing fields, weeding, etc. Yet, most agricultural workers have various sources of income (e.g., agriculture on their own homestead, quarry work, petty trade).

Except for the security of dwelling sites, the land reform did not offer as much to agricultural workers as to former tenants. Pleading poverty themselves, the new owner-cultivators were also reluctant to increase agricultural wages and to share their gains from the land reform. Consequently, conflicts emerged between the agricultural workers and the new smallholders, who had been allies in the struggle against landlordism. These conflicts eventually led to the formulation of the Kerala Agricultural Workers Act in 1974, which has changed labor relations significantly. Provisions of this act include (see Franke & Chasin 1994: 85-86):
- guaranteed permanency of employment for attached laborers;
- fixed and reduced working hours;
- revised minimum wage;
- provident fund for death payments;
- district-level arbitration board;
- unemployment insurance (since the early 1980s);
- pension scheme for agricultural workers (since the early 1980s).

The Kerala Workers Act and the related welfare schemes have eliminated semi-feudal, patron-client and serf-like relations, and mitigated many fears arising from such things as the denial of employment or eviction from the dwelling site. The reform has generally rendered agrarian relations less personal, more capitalistic and more formally regulated. However, agricultural workers have remained one of the poorest and most vulnerable groups of Kerala's society (Herring 1989: 99-104; Franke & Chasin 1994: 85-86).

Labor unionism is widespread and very influential in Kerala. The power of labor unions is particularly strong among loaders and porters. Unions allow only a limited number of members per local working unit. Other people are prevented from loading and unloading in their "territory." If a member retires, he or she is usually replaced by one of his or her family members. In this way, the labor unions (including those affiliated with secular political parties) reinforce the labor-market segmentation along caste lines (Törnquist 1995: 30). Politically affiliated labor unions of the industrial and plantation sector have also spread to agricultural laborers. The numerous local units of Kerala's unions regulate labor markets to a significant extent. They reserve jobs to locals (as opposed to people from other regions and states, where wage rates are lower) and also to people of particular families, castes or political affiliations. In the case of jobs requiring special skills such as coconut-tree climbing, for example, labor unions restrict the access to people from traditional coconut-tree-climber families. For other agricultural work, labor unions usually do not restrict access. Instead, they organize agricultural workers to bargain for higher wages from cultivators. Consequently, daily agricultural wages have increased significantly, particularly since the mid-1980s (see Fig. 7).

In response to labor-market regulations and fairly high wages, many cultivators attempt to save hired labor by shifting to less labor-intensive cropping patterns and cultivation methods. Consequently, available days of employment for agricultural workers have declined. Between 1963-64 and 1983-84, for example, the average number of working days per year declined from 198 to 147 for males, and from 165 to 115 for females (Directorate of Economics and Statistics 1985, cited in Franke & Chasin 1994: 89). Likewise, the proportion of permanent workers decreased in relation to casual workers, from 28 percent in 1972-73 to 23 percent in 1982-83 (Vaidyanathan 1986, cited in Oommen 1993: 17). Reduced employment opportunities, especially in rice-growing areas, in turn, has forced agricultural workers to demand higher wages. As a consequence, a vicious circle is created by the cultivators' labor-saving strategy and the agricultural workers' demand for higher wages.

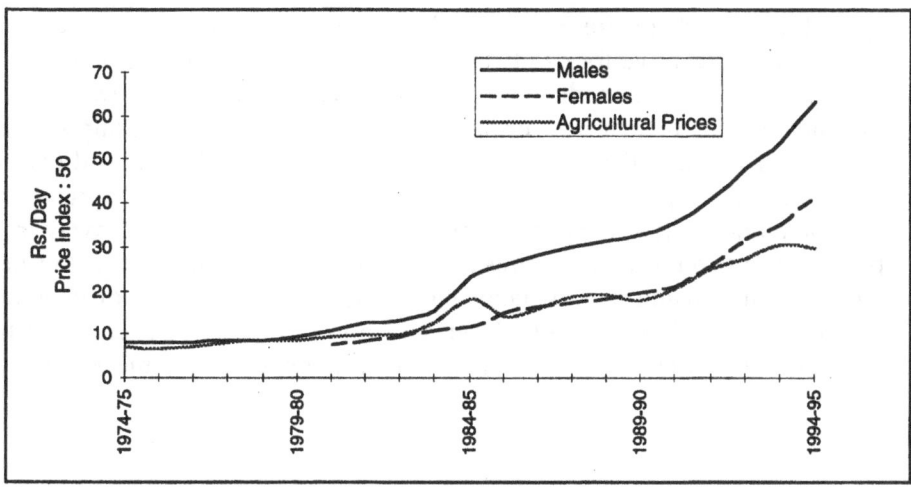

Fig. 7   Daily Average Wage of Paddy Workers in Kerala
Source: Directorate of Economics and Statistics, Season & Crop Report, various issues.

More recently, however, agricultural labor seems to have become scarce in particular regions. This may appear paradoxical in times of high and growing unemployment in Kerala. However, also low-class people prefer less toilsome and less filthy jobs (e.g., loading, driving auto-rikshaws or trucks, petty trading) to agricultural work. Furthermore, the children of traditional agricultural-worker families are now formally educated. Some of them want to move up socially and therefore choose professions independent of their inherited status (caste). Generally, the young formally educated people prefer to stay unemployed and wait for a favorable job offer in the tertiary sector rather than work in agriculture or become self-employment (see Mathew 1995). (The fact that families can *afford* to have unemployed young grown-ups may be partly ascribed to the influx of Gulf money. Verifying such a hypothesis, however, is beyond the scope of this study.) The declining importance of agricultural work for wage laborers is also reflected in their recent indifference to agricultural mechanization: In the 1970s and 1980s, wage laborers still militantly opposed mechanization. In sum, changing job preferences and alternative (though limited) employment opportunities in the tertiary sector have led to an agricultural-labor scarcity in peak seasons and to a better bargaining position for agricultural workers.

The new labor relations as well as the relatively high wage rate have influenced recent changes in cropping patterns. The new condition that agricultural workers are not on call around the clock as before, is a particular disincentive for paddy cultivation, which needs intensive care and is very sensitive to immediate completion of certain operations (e.g., repair of embankment breaches). Furthermore, paddy cultivation requires much extra labor at particular seasons (e.g., for harvesting). The reduced control over workers, regulated working hours, general scarcity of agricultural labor and high wages, therefore, contribute especially to the shift away from paddy cultivation to labor-saving crops such as coconut or rubber. Furthermore, the increasing cost of labor encourages the application of chemical herbicides as well as agricultural mechanization, which, however, is limited to regions of extensive plains (e.g., Palghat) for technical and ecological reasons.

5.4.3 Spread of Formal Agricultural Credit

For many cultivators in Kerala, capital has become an important input for cultivation. Since the nationalization of banks in 1969 and the resulting spread of bank offices into "rurban" areas of Kerala (according to the CMIE 1994, cited in State Planning Board, Economic Review 1995: S182, the number of bank offices has increased from about 600 in 1969 to almost 3,000 in 1995), the role of informal moneylenders in providing agricultural credit has declined significantly. The per-acre availability of agricultural credit from banks is far higher than elsewhere in India. Even compared with agriculturally advanced states like Punjab or Haryana, the amount of agricultural credit available per acre is more than double in Kerala. The main suppliers of formal agricultural loans are cooperative societies and public-sector commercial banks, both of which get backing from the National Bank for Agricultural and Rural Development (NABARD) (Sunanda 1991: 155-156). Short-term agricultural loans with a repayment period of 18 months (crop loans) are given for agricultural inputs such as seeds, fertilizers and pesticides, and to meet overhead costs for labor, irrigation and electricity. Cooperative Primary Agricultural Credit Societies disburse the crop loans both in cash and in kind (seeds, fertilizers). Some of these cooperatives also engage in product marketing and agro-processing. Medium-term agricultural loans (repayment period of five to seven years) and long-term agricultural loans (repayment period up to 15 years) are for long-term investments in plantations with long gestation periods, irrigation infrastructure, land development, farm mechanization, livestock, etc. (Sunanda 1991: 172-175).

D.K. Desai maintained that Kerala was the only state where the agricultural-credit needs of farmers are fully met by cooperative and commercial banks. In 1984-85, the supply of formal agricultural credit was 102 percent of the demand (Desai 1988, cited in Oommen 1993: 12). But a more detailed analysis revealed that in the mid-1980s, only about one third of smallholders (i.e., those with landholdings below five acres) took formal agricultural loans, whereas medium and large landowners (above 10 acres) took several formal agricultural loans from more than one bank (Sunanda 1991: 159-160). Thus, smallholders in particular still rely on their own savings and, especially, on informal moneylenders, whose services involve less formalities but higher interest rates than bank loans. For agricultural investments, a few farmers also prefer ordinary bank loans to agricultural loans, which involve more formalities. This group of farmers includes lessees who do not qualify for agricultural loans because formal agricultural loans are restricted to cultivation on one's own land. On the other hand, a high portion of agricultural loans seems to be used for non-agricultural purposes, including consumption (Sunanda 1991: 162-163).

Although the availability of agricultural credit has improved significantly and is relatively good in Kerala, and although most farmers have thus become autonomous not only from landlords but also from usurers, farmers do not necessarily engage in more intensive agricultural practice (as neo-Marxist development theory suggested). Rather, the availability of long-term investment loans, which usually are not given by informal money-lenders, facilitates the farmers' shift to plantation crops with a long gestation period. Furthermore, long-term loans are important for environmental improvements (e.g., soil conservation measures). However, the share of agricultural credit used for land development is only about 2 percent (NABARD, cited in P.S. George 1995: 19).

The general availability of agricultural inputs such as seeds, chemical fertilizers, pesticides means that scarcity of inputs is also not a major obstacle to Kerala's agricultural development. The supply of particular seeds and seedlings, however, may be insufficient and still dependent on distribution by state-government farms (as opposed to private nurseries). Furthermore, after the removal of the substantial central-government subsidies on chemical fertilizers in 1992 under the New Economic Policy, farmers in Kerala initially used less chemical fertilizers (State Planning Board, Economic Review 1995: 49). However, it is beyond the scope of this study to assess whether the reduced use of chemical fertilizers led to insufficient replacement of soil nutrients and soil degradation, or whether it was accompanied by more efficient and environment-friendly fertilizer use. Also, fertilizer subsidies were reintroduced in the election year of 1996.

5.4.4 Infrastructure and Technology Development

The state government of Kerala plays a major role with respect to agricultural infrastructure. The main area of government support has been the development of irrigation. No less than 13 to 22 percent of the total plan outlay in Kerala's various Five-Year Plans between 1951 and 1997 has been allotted for irrigation. By 1995, the state government and private parties had brought nearly one million acres of land under irrigation (10 percent of Kerala's total area) (State Planning Board, Economic Review 1995: 68-69). In line with the Green-Revolution strategy, the state irrigation schemes focused on large-scale canal irrigation and on irrigating paddy.

However, the implementation of the expensive large-scale irrigation projects was not efficient, their operation has been ineffective and their maintenance has been poor. Despite their focus on paddy, the government irrigation schemes were not able to prevent paddy conversions. Only recently did the state government start to pay increased attention to groundwater development, minor irrigation projects with community participation, and the irrigation of dryland crops. These new small irrigation schemes seem more promising for sustainable agricultural development in Kerala than large irrigation schemes (Jacob 1996).

In regard to soil and water conservation, which are of particular importance for Kerala's environmental sustainability, the state government has made initial efforts. So far, however, Kerala's Soil Conservation Department has covered only an area of 325,000 acres out of an estimated 3.75 million acres vulnerable to soil erosion, and the measures of the Soil Conservation Department can be questioned from a technical point of view (see Narayanan 1996: 13). The Centrally Sponsored National Watershed Development Programme, initiated in 1992, also aims at "developing" about 138,000 acres in 114 watersheds of Kerala (P.S. George 1995: 10-11). But government-sponsored water and soil conservation measures have not been able yet to substantially prevent soil erosion and landslides in hilly tracts that were colonized a few decades ago. In these step highland areas, the traditional *kayala* system, which is usually practiced on gentle midland slopes, also seems inappropriate. *(Kayala* stands for terracing slopes with mudwalls on which reeds are grown. For decades if not centuries, this agricultural practice may have effectively prevented soil erosion in Kerala's midland.)

For the development and extension of agricultural technology, India's Council of Agricultural Research, Kerala's Department of Agriculture and specialized Commodity Boards are of major importance. The Kerala Agricultural University (KAU), a unit of the Department of Agriculture, runs five regional agricultural-research stations and about a dozen specialized research stations. In these stations, cultivation methods, plant-protection measures and new varieties of crops (rice, coconut, cashew, pepper, cardamom, banana, pineapple, vegetables, aromatic and medicinal plants) are developed, tested and adapted to the agro-climatic conditions of Kerala. Since 1982, the KAU has also undertaken farming-system research for coconut-based, tuber-based and rice-based farming systems, for integrated crop-livestock-fish farming, and for homestead farming. In addition, specialized Commodity Boards have been established for rubber, spices, coffee and tea. These Commodity Boards engage in research, run extension units and give marketing assistance. Particularly the Rubber Board, based in Kottayam as a subordinate of India's Ministry of Commerce, has the reputation of being efficient, corruption-free and comprehensive in the services it offers. The Rubber Board extends technical advise and training through its field offices, which are spread all over Kerala, and also provides substantial subsidies and loans for planting and replanting, as well as marketing support. Its appropriate and efficient technology development and extension work is a significant incentive for farmers to begin rubber cultivation.

For most crops, agricultural extension is organized through Kerala's State Department of Agriculture. In Kerala, government-sponsored agricultural extension started in 1952 through community-development blocks. Since then, the agricultural-extension system has been modified several times. In 1987, *Krishi Bhavans* (Agricultural Offices) were established in each panchayat in a move to streamline and decentralize the Department of Agriculture. Each Krishi Bhavan consists of one agricultural officer, two or three agricultural demonstrators and an advisory committee representing bureaucrats, cultivators and agricultural laborers. Krishi Bhavans disseminate new seed varieties and cultivation methods and they advise individual farmers or groups of farmers during farm visits. The technical support is usually based on recommendations developed by the KAU and published in the *Package of Practices Recommendations*. Furthermore, Krishi Bhavans supply agricultural inputs (e.g., seeds, fertilizers), and support farmers with advice on small-scale irrigation, soil conservation and agricultural marketing. They also offer agricultural credit. In most cases, these services are related to government-sponsored agricultural-development schemes, for which the Krishi Bhavans are the main implementing organizations.

This decentralized agricultural-extension system has the advantage that farmers get comprehensive service and access to various state-level agricultural-development schemes through only one local organization. Because they are understaffed, however, the Krishi Bhavans cannot pay equal attention to all of the many programs. Consequently, they focus on paddy cultivation and, depending on the region, on one or two other crop-specific schemes. Moreover, it is impossible for the agricultural officer and his or her staff to regularly visit and advise all of the thousands of cultivators in a panchayat individually. Nor has the decentralized agricultural-extension system generally led to intended bottom-up and participatory agricultural planning. Rather, it has made for greater efficiency in the implementation of the Department of Agriculture's schemes. For instance, one agricultural officer of a local Krishi Bhavan reported that she once made suggestions to modify a state-level scheme to local conditions. Although the Department of Agriculture encourages agricultural officers to make such recommendations, two years later, she has not heard anything more about the matter. Another problem is that agricultural officers are frequently transferred – often as a consequence of a change of the state government. This gives the officer too little time to understand local-specific problems and to build-up confidence with the cultivators of the respective panchayat.

The technology development and extension programs of the state of Kerala contribute to increased productivity of particular crops. However, the Green-Revolution approach was not as successful as in Punjab or Gujarat, where grain production responded better to irrigation than in Kerala. Furthermore, although state interventions have focused on paddy cultivation, they have not been able to prevent the paddy conversions in Kerala. Since the late 1980s, particularly with the group-farming program, which encourages joint farm management among smallholders of the same micro-watershed, the Department of Agriculture shifted from a primarily technocratic approach to a more institutional approach. Although the productivity of paddy increased in many groups in the initial years after 1989, the group-farming program failed in most cases because (part-time) farmers showed no interest and lacked the time for joint farm management. Often, the only common interest among farmers was to receive subsidized farm inputs (e.g., chemical fertilizers, mechanical tillers and tractors) through the group. When subsidies for group farming were withdrawn in the early 1990s, most groups became inoperative (Törnquist 1995: 59-65). Recently, the Department of Agricultura also began to disseminate eco-friendly technologies (e.g., bio-fertilizers and bio-pesticides) and cultivation methods (e.g., reduced fertilizer application, integrated pest management).

The current budgetary constraints of the state of Kerala may lead to disinvestment in crucial areas such as agricultural technology and infrastructure, where the government has played a central role. This trend, however, may also have positive attributes. It may force the state government to further reorient its infrastructure policies away from expensive large-scale irrigation projects that have limited impact on Kerala's agricultural development, and toward projects with big potential such as minor-irrigation or rain-fed agriculture. Furthermore, the downsizing of centralized planning at national and state level can open up scope for local-level planning under the revised Panchayati Raj Act of 1994 and lead to the better involvement of non-governmental organizations and people's participation.

5.4.5 Development, Organization and Regulation of Crop Markets

Long-distance trade in agricultural commodities from Kerala may have already started about five millenniums ago. Market-oriented agricultural production became substantial in Kerala in the 19th century, with the establishment of the plantation economy under direct and indirect British rule. Apart from spices, coffee, tea and rubber became important export crops, and plantations were established on former forest land in the hill tracts (Prakash 1987). In the early 20th century, a cash-crop boom and further commercialization of agriculture encouraged the colonization of vast areas in hilly regions (Tharamangalam 1984). Due to the fairly strong market orientation of Kerala's agriculture, rice has had to be imported since the medieval period. Smallholders, too, produced for the market. In North Malabar, for example, many smallholders were engaged in market-oriented agriculture of pepper and coconut and neglected subsistence production of paddy (see Menon 1994: 21-24).

Despite the *Grow More Food Campaign* of the 1950s and the sustained efforts of Kerala's government to promote paddy cultivation, the shift from food crops to cash crops has continued and even accelerated. Apart from traditional export and cash crops (e.g., pepper, coffee, rubber, coconut, cashew), new cash crops (e.g., fruit, vegetables, medicinal plants) and partly also "classic" food crops (paddy, tapioca) are produced for domestic and local markets. Approximately half of the gross output of Kerala's agriculture is sold in other states of India or exported to foreign countries. Interstate trade has become more important than international trade in the post-Independence period. By 1980, the share of international trade had declined to less than 25 percent of total trade from Kerala (Isaac 1994: 368, 389).

Because of India's export regulations, most agricultural products appear only on domestic markets. Furthermore, domestic demand for unprocessed and processed agricultural products is increasing – thanks to a growing non-agricultural population. Monetization has also rendered agriculture more market-oriented. However, subsistence production is still important in Kerala. Nearly every household grows trees, vegetables and fruit in its home garden, and uses the many different products for home consumption. Particularly for poor households, subsistence production forms a substantial source of food, fuel, wood and fodder (see Jose 1991).

International markets are still the domain of a few state-promoted products such as cashew, pepper, cardamom, ginger, coffee and tea (Isaac 1994). However, India's participation in the World Trade Organization and trade liberalization under the New Economic Policy may open up export possibilities for animal-husbandry products, cut flowers and processed and non-processed fruit and vegetables – products, for which, according India's Trade Development Authority, Kerala has comparative advantage.

In Kerala, trade in agricultural commodities is mainly carried out by private traders (local petty traders, country buyers, commission agents, wholesalers, etc.). Merchants who operate from terminal markets using agents and sub-agents control many crop markets in Kerala (P.S. George 1995: 15). The traders may or may not exploit the farmers. Oommen (1993: 13), for example, maintained that "moneylenders and traders as an exploitative class appropriating the surplus in the agricultural sector do not seem to be dominant in the State." Rai (1988), on the other hand, noted that Kerala's traders and the many middlemen appropriate an exploitative and unjustified, high margin between the farm-gate and retail prices.

Apart from different perspective and ideology, contrary views about traders may also reflect particular experiences with different products, markets and regions. In general, it seems that traders in Kerala do not excessively exploit cultivators. Unlike in other states of India, crop markets are not organized by all-powerful landlords-cum-moneylenders-cum-traders. Even before the abolition of landlordism in 1970, small cultivators seem to have sold gardenland products to specialized traders and not to their landlords (see Menon 1994: 24). With the improved availability of agricultural credit from banks, furthermore, cultivators have become less dependent on cash advances from traders. Consequently, crop markets are not strongly interlocked with land and capital. Furthermore, there has always been a culturally deep-rooted suspicion toward traders and business-men that is reflected in the Malayalam proverb *"kachavadom* (trade) is *kachakapadam* (dressed-up deceit)"* (Zachariah & Sooryamoorthy 1994: 38). This cultural attitude, and the high level of education, may also help to check the traders' activities.

Apart from private parties, cooperatives are also involved in marketing agricultural products. Numerous panchayat-level cooperatives (Primary Agricultural Credit Societies, Primary Marketing Societies, etc.) sell a variety of agricultural products through state-sponsored apex cooperatives such as KERAFED (copra, i.e., dried coconut meat), the Marketing Federation (copra, cashew, arecanut, pepper and other spices), CAPEX (cashew), the Rubber Marketing Federation (rubber) or CAMPCO (arecanut, cocoa). However, cooperatives in Kerala hold only a small share in marketing and agro-processing and they have no effective control over market forces (State Planning Board, Economic Review 1995: 62). Moreover, the state-subsidized cooperatives are often mismanaged, uneconomical and tainted with party politics and favoritism. On the local level many of them are "battle fields for petty politicians and communal leaders" (Suresh & Molly 1990: 49).

The state of Kerala also engages directly in crop markets. The state government has introduced occasional state-monopoly procurement of cashew as well as *regulated markets* for coconut and arecanut in the Malabar region. India's government has introduced 6,700 regulated markets all over the country in order to ensure fair prices, reduce marketing charges and eliminate unhealthy market practices. A market committee consisting of representatives of the state government, legal bodies, traders, commission agents and cultivators fixes market charges and prices and supervises the marketing practices (Datt & Sundharam 1995: 493). India's central govern-ment also regulates imports of many agricultural products. Grains, edible oils and rubber, for example, are not allowed to be imported or are subject to tight import quotas that are allotted to government agencies. For many other agricultural commodities, big import duties exist (Randhawa 1994: 357). Important cash crops of Kerala's farmers such as rubber and coconut enjoy massive government protection. Under the New Economic Policy, however, the central government relaxed import restrictions on edible oils in 1995, in spite of protests from coconut-oil millers in Mumbai (Bombay). Consequently, farm-gate prices for coconut and copra in Kerala declined substantially. In the case of rubber, on the other hand, the rubber-grower lobby groups have successfully prevented import liberalization despite skyrocketing rubber prices in 1995, raw-material shortages and demands by rubber-processing factories to import cheaper rubber from abroad (The Hindu, various issues; Indian Express, various issues).

Prices of cash crops, particularly of export crops, generally show wide seasonal and annual fluctuations because they are usually determined outside Kerala by important consumers and agro-processing industries and because hardly any buffer stocks are available in Kerala. In order to ensure remunerative farm-gate prices, the state government sometimes introduces support- or fixed prices for particular products such as copra or raw cashewnut. However, India's New Economic Policy may soon include the withdrawal of administered prices and food subsidies. This would probably lead to higher price fluctuations and increased cultivation costs. Withdrawal of food subsidies is likely to result in higher paddy prices, but paddy cultivators in Kerala would still face the competition of rice imports from other states, where paddy can be produced more cheaply.

Generally, farm-output relations and agricultural-product markets are highly crop- and region-specific. Therefore, no universal conclusions about the nature and functioning of crop markets in Kerala can be made. However, marketing opportunities in Kerala seem relatively good, thanks to the general availability of traders and established marketing networks. Marketing information is accessible through newspapers, which are read daily by most people in Kerala. Furthermore, the road network is well developed and enough vehicles are available to transport agricultural products. Storage facilities, post-harvest handling and local agro-processing, on the other hand, are still deficient and may constitute a bottleneck for agricultural development in Kerala (Rai 1988; Mohandas 1995).

A highly market-oriented agriculture suggests that prices play a dominant role in agricultural development. In general, prices for agricultural products have increased during the past decades in Kerala but, as the parity index (ratio of prices received by farmers to prices paid by farmers) shows, cultivation costs have increased at a similar or even higher rate. Sector-specific policies and subsidies (e.g., import substitution and industrial protection) may be discriminatory against agriculture because they raise prices for agricultural inputs and overvalue exchange rates, thus worsening export potentials for agricultural products (Bhalla 1994: 5). However, it would be premature to conclude that a deteriorating price-cost ratio has discouraged farmers from engaging in intensive agricultural production or, on the other hand, that structural adjustment and resulting higher agricultural prices would automatically lead to agricultural growth.

Probably, *relative prices* of agricultural commodities have influenced cultivators' decisions more than overall agricultural price trends, particularly regarding crop selection. Among many other factors, stagnant rice prices influence cultivators to shift away from paddy, and increasing rubber prices motivate them to start new rubber plantations. However, in other cases (e.g., cashew; see Ch. 7), price changes do not have as much influence as neoclassical theory would predict. Not only prices but also the marketability of particular agricultural products needs to be considered. Particularly for new cash crops such fruit, vegetables or flowers, marketing facilities and structures are not fully established yet, and marketing opportunities vary from region to region (see case study on pineapple in Ch. 6).

5.4.6 *Relative Importance of Socioeconomic and Technical-Material Factors*

This discussion of recent changes in Kerala's agricultural relations suggests that today's cultivators operate in a context where the system of private landownership, labor markets, capital markets and crop markets are fairly well developed. Furthermore, agricultural markets are generally not strongly interlocked. For instance, landowners cannot link the access to land with credit or marketing, because landlordism and tenancy have been abolished. Furthermore, because agricultural credit is available from banks, moneylenders cannot give credit at exorbitant interest rates and traders cannot give pre-harvest payments under terms very unfavorable to the cultivator. In other words, unlike in many other states of India, the exploiting landlord-cum-moneylender-cum-trader nexus, discussed by neo-Marxist theory, do not apply in Kerala. This, however, does not mean that the respective factor and output markets are "perfect" in a neoclassical sense. Rather, state regulations, lobby groups such as labor unions, and informal rules influence agricultural markets, including the level of transaction costs. The marketing infrastructure within Kerala is fairly well-developed (e.g., village marketplaces, roads, vehicles), but prices, formed mostly in national and global markets, tend to fluctuate. In general, crop markets seem fairly "open" (i.e., working according to the principle of demand and supply). This is partly because of strong competition among the many private and cooperative traders. This general overview, however, cannot provide information on instances of forced commerce and different involvement of various classes of cultivators into agricultural markets.

Agrarian relations in Kerala are unique and create specific conditions that influence and structure the decisions and practices of cultivators. Some aspects of these agrarian relations contribute to agricultural stagnation, and to changes in cropping patterns and altered cultivation methods, some of which are not sustainable either environmentally, economically or socially. However, conclusions regarding the relative importance of the various socioeconomic and technical-material factors for sustainable development are difficult to arrive at on the basis of a general overview. Moreover, these socioeconomic conditions for farming are changing; and change does not come about automatically but rather through interaction and negotiation between rival interest groups. In order to secure the Land Reform Act and to get it implemented effectively, for example, tenants, agricultural workers and communist politicians struggled for more than two decades against landlords (see Herring 1983153-216; Franke & Chasin 1994: 81-86). After 1970, conflicts between agricultural workers and the newly-landed small-holders led to formalized agricultural-labor relations as detailed in the Kerala Agricultural Workers Act of 1974 (see Herring 1989). Bargaining of agricultural workers and trade unions for higher wages continues to the present.

When it comes to the application of agricultural technology, the farmers do not simply adopt the practices advocated by government agencies. Interactions among cultivators, as well as experiments by the cultivators themselves, need to be considered in addition to the government-sponsored development and extension programs. Cultivators adopt new seeds or cultivation methods selectively and adapt them to their own circumstances. Also, markets do not just emerge automatically or through an "invisible hand" but are created by traders, agro-processors, state agencies and consumers. Such interactions and processes, particularly the "making" of markets, are given fuller attention in the following two case studies.

6 Pineapple Cultivation in Vazhakulam

6.1 General Patterns of Pineapple Production and Trade

Pineapple is grown in tropical and sub-tropical regions throughout the world. Over the past 25 years, global production has more than doubled; in 1995, it reached about 11.5 million tonnes (FAO 1990-1996). At present, the major pineapple-producing countries are Thailand, the Philippines, Brazil, India, Nigeria, China and Taiwan. Thailand and the Philippines are not only the main pineapple-growing but also the main pineapple-exporting countries. In many other countries, a high proportion of pineapple is sold in domestic markets: only about 15 percent of the pineapple produced appears on the international market. Yet, global pineapple exports increased from 0.61 million tonnes in 1970 to 1.75 million tonnes in 1995 (FAO 1990-1996). For the past 25 years, the international trade in fresh pineapple has increased faster than the trade in canned pineapple, but remains smaller (0.77 million tonnes compared with 0.98 million tonnes). The increased trade in fresh pineapple is not only a consequence of improved transportation technology but also caused by changing consumer preferences that, at least in Western countries, have shifted to fresh-food products.

In India, only after Independence in 1947 did the production of pineapple, which was introduced in 1548 by Portuguese colonialists, start to grow substantially; between 1965 and 1995, for example, it grew rapidly from 280,000 tonnes to 820,000 tonnes. Pineapple exports, on the other hand, remained negligible. In 1995, for example, only 150 tonnes of fresh pineapple were exported from India (FAO 1990-1996). The rapid growth of pineapple production has been influenced by the overall population growth and, in particular, by increased demand from a swelling urban middle-class, which does not grow pineapple but rather buys it in the market.

In India, the total area under pineapple cultivation is estimated to be 210,000 acres (Becker 1988, cited in S.K. Sen 1990: 259). Generally, India's pineapple plantations are much smaller than those in Thailand, the Philippines or Hawaii. Furthermore, pineapple cultivation is not mechanized and generally less intensive. The most common pineapple variety grown in India is the *Kew* pineapple (Cayenne group). The relatively big fruit of Kew pineapple is consumed as fresh fruit or is processed into canned pineapple slices, juice, jam and syrup. The other variety grown in India, *Mauritius* pineapple (Queen group), is sweeter but smaller than Kew pineapple. Mauritius pineapple is not used in fruit-processing factories. Due to its superior taste, however, it is very popular as fresh fruit. To my knowledge, Kerala is the only region in India where Mauritius pineapple is grown commercially.

The first pineapple-processing industries in India started only after Independence in 1947. They have mainly produced for the domestic market. From the 1960s to the 1980s, the industry expanded in the vicinity of the main pineapple-growing areas. In the mid-1980s, however, the Indian pineapple industry met a crisis. Many pineapple-processing factories closed down. Others shifted from pineapple canning to the production of mango juice, pickles and other fruit products. According to the interviewed pineapple processors in Kerala, the reasons for this were:
- increased production costs (particularly for tin cans);
- preference of consumers for the cheaper and tastier fresh pineapple;
- increased availability of fresh fruit in urban markets due to improvements in transportation;
- the resulting shortage of raw material for processing;
- lack of export opportunities due to strong international competition and high international quality standards.

The Indian government has supported the pineapple sector to a limited extent since the 1960s, for example, by disseminating processing technologies and introducing new varieties of the fruit; but it has not regarded pineapple as a core area for agricultural-development assistance. Therefore, pineapple production, marketing and processing are generally based on private initiatives.

In Kerala, too, pineapple cultivation has grown since Independence. Based on information from pineapple wholesalers, I estimate that over 100,000 tonnes pineapple are produced there annually. However, extent and trends of pineapple cultivation differ widely among Kerala's districts. In most regions, pineapple cultivation is insignificant, limited to homesteads and declining. From the 1950s to the 1980s, however, commercial Kew-pineapple cultivation was widespread in Thrissur District and also in Kannur District. In Thrissur District, large-scale pure pineapple plantations existed on upland. In Kannur District, pineapple was often intercropped on the widespread cashew plantations. In both cases, pineapple was grown to supply Kerala's fruit-processing industry. In the 1980s, pineapple cultivation in Thrissur and Kannur District declined sharply because of the factories' falling demand (see above) and because of decreasing yields of pure pineapple cultivation on upland.

By contrast, Mauritius-pineapple production in Vazhakulam, Ernakulam District, has been booming since 1983. That region has become the most important pineapple-producing area in Kerala. Vazhakulam has also become Kerala's most important pineapple-assembly market, and has gained national importance, too. Nowadays, the intensive, commercial cultivation of Mauritius pineapple is also spreading to neighboring areas in Ernakulam, Kottayam and Idukki Districts.

6.2 Local Pineapple Boom

This case study concentrates on the region around the village of Vazhakulam in south-east of Ernakulam District. This region – in the following simply referred to as Vazhakulam – comprises four panchayats, covering an area of about 100 square kilometers and having a population of nearly 60,000, in 15,000 households. Agriculture – particularly rubber production and further also pineapple production – is the mainstay of the economy there. For many households, however, agriculture has become a subsidiary economic activity because of other income opportunities and the declining size of operational landholdings.

Vazhakulam belongs to Kerala's midland. The topography is dominated by gentle hills which are intercepted with valleys. The current vegetation is almost completely formed by intensive, market-oriented agriculture (rubber plantations, coconut groves, paddies, etc.). Almost all cultivable land is put under agriculture. Present land-use changes relate to changes within agriculture (changes in cropping pattern) or to the shift to non-agricultural uses (rurbanization). As elsewhere in Kerala, the most predominant recent change has been the conversion of wetland paddy into dryland. In Vazhakulam, unlike elsewhere in Kerala, the predominant change related to wetland conversions is the shift to pineapple cultivation.

The beginnings of pineapple cultivation in Vazhakulam go back to the colonization of the region in the last century. Since then, gentle hills have been terraced with mudwalls on which pineapple and reeds are grown *(kayala* system). The kayala system prevents soil erosion; pineapple plants, with their fibrous root network, help to preserve the terraces. The thorny leaves of Mauritius pineapple also formed natural fences that mark and protect property. Until the 1960s, the most widespread pineapple variety was Mauritius, but Kew pineapple was also grown. Yet, the cultivation of pineapple was not substantial. Although the fruit also served as cattle fodder and as compost, the local people hardly ever ate pineapple, which they believed to contain ingredients that are harmful to the human body. In the 1960s and 1970s, this misbelief vanished. Still today, pregnant women in Vazhakulam do not eat pineapple because the locals believe that pineapple has an abortive character. Nor is pineapple given to pregnant cows. Scientific confirmation of the abortive character of pineapple could not be traced in the literature, but an early source (Volkamer 1987 [1714]) describes how prostitutes in India made use of the abortive character of pineapple in the 18th century.

Early but very limited sale of pineapple from Vazhakulam began in the late 1950s. On infrequent journeys to the marketplace in Aluva, which lies about 40 km to the north-west, people from Vazhakulam occasionally took two or three pineapple fruits to sell in the marketplace. In the mid-1960s, agents of fruit-processing factories from the town of Thrissur, which is about 90 kilometers away, started to come to the weekly village market in Vazhakulam to buy pineapple (Kew variety). In the late 1960s, the first fruit-processing unit in Vazhakulam started production (at the peak in the early 1980s, there were six small-scale fruit-processing factories in this region). Pineapple from there was very popular among agro-processors because of its good quality. During the 1970s, the trade in pineapple grew steadily, and local traders got increasingly engaged in pineapple trade. They sold Kew pineapple to fruit-processing factories in the towns of Thrissur and Kottayam and, since the late 1970s, also to factories in the cities of Chennai (Madras) in Tamil Nadu, and Bangalore in Karnataka. Although the cultivation of pineapple became market-oriented, it generally remained a small side-business for both traders and cultivators at that point.

Major changes took place in the early 1980s:
- new cultivation technologies, including artificial floral induction, the use of chemical fertilizers and irrigation, were introduced in 1980;
- a pineapple-specific marketplace was built by the Panchayat in 1982;
- production, as well as fresh-fruit marketing, of Mauritius pineapple (as opposed to Kew pineapple) became popular in 1983.

Eventually, fresh-fruit pineapple business started to boom in Vazhakulam and became important for many cultivators and traders. Pineapple production there has increased at least tenfold since 1983.

The first phase of the increase was due to pineapple cultivation on upland – especially temporary intercropping on young rubber replantations.

However, this cultivation system reached a saturation point; practically all rubber replantations in Vazhakulam are used for pineapple intercropping. Since the late 1980s, the growth in pineapple production has come from pure pineapple cultivation on former paddies, which are converted into dryland for the purpose of growing pineapple (although the conversion of paddies is illegal under Kerala's Land Utilisation Act of 1967). To a smaller extent, pineapple is also grown on upland, both as an intercrop and as a pure crop. According to my own estimate, about 3,000 to 3,500 acres in Vazhakulam (20 to 23 percent of the cropped area; 12 to 14 percent of the total area) were, at least partly, under pineapple cultivation in 1995. Unlike in other areas of Kerala, where Kew variety is more widespread, more than 95 percent of the pineapple grown in Vazhakulam is of the Mauritius variety. Nearly all households grow a few pineapple on the homestead and about 800 to 1,500 households (5 to 10 percent), predominately middle-class and rich farmers, as well as leaseholders, are also engaged in commercial pineapple cultivation. At present, however, pineapple cultivation in that region is not expanding rapidly anymore: most suitable land is already under pineapple cultivation and competition from elsewhere is growing. Currently, the highest growth rates of pineapple cultivation are in adjacent areas of Ernakulam, Kottayam and Idukki Districts.

Apart from the expansion of pineapple cultivation, more intensive cultivation techniques have become widespread since the early 1980s. Most important is the application of the flower-inducing chemical *ethephon*. Pineapple plants flower 25 days after the application of this plant-growth regulator; another 95 to 110 days later, the fruit is ready to be harvested. Because of these properties of ethephon, cultivators are able to time the harvest of entire pineapple fields, and to break the dependence on natural seasons. Main harvests can be scheduled for any time of the year, provided the growing fruits get sufficient water. Farmers usually divide the pineapple field into two to four parts, where plants are given the flower-inducing chemical at different times. In this way, pineapple growers get about two to four smaller yields per year instead of one very big yield.

Parallel to the expansion and intensification of pineapple cultivation, has been the growth of pineapple marketing from Vazhakulam. In 1983, the sale of Kew pineapple to fruit-processing factories was largely replaced with the sale of Mauritius pineapple to fresh-fruit wholesalers. Since then, sales of pineapple (Mauritius variety) have increased every year. Apart from urban markets in Kerala, terminal markets in other states (first Bangalore and Chennai; since the late 1980s also Mumbai (Bombay), Hyderabad, Ahmedabad, etc.) have gained importance. Today, about 25 to 50 percent of the pineapple sales from Vazhakulam go outside Kerala. Mumbai has become the single most important market for this pineapple. Very small amounts are exported by air to Gulf countries via agents based in Thiruvananthapuram and Alappuzha. For many local traders, pineapple marketing has become a daily and very important business. The number of pineapple traders has increased from a handful in the early 1980s to more than 200 in 1995, 50 of which are large wholesale merchants. Nowadays, about 200 to 300 tonnes of pineapple produced both in Vazhakulam and in new adjacent pineapple-growing areas go through the Vazhakulam market daily.

At the same time, the processing of pineapple has declined in Vazhakulam. Three of six fruit-processing factories have survived, but have substituted other fruit products for pineapple-based product. The Kerala Horticultural Development Programme, a joint venture between the Commission of European Communities and the Kerala State Department of Horticulture, however, is planning to start a large fruit-processing factory in that region in November 1997.

6.3 Socioeconomic and Environmental Impact

For 5 to 10 percent of the households, commercial pineapple cultivation provides an important source of income. Middle-class and rich people are particularly able to make use of the remunerative commercial pineapple production and marketing opportunities. However, the income from pineapple production is quite unsteady due to price fluctuations in the open market. Cultivators have to sell at any price because pineapple is a perishable fruit that cannot be stored like food grains, for example. Poor peasants, furthermore, find it difficult to start large-scale pineapple cultivation because this requires many resources, substantial investments and relatively high production costs.

However, also for poor households, the massive growth of pineapple cultivation and marketing in Vazhakulam has had positive effects. The pineapple boom has created additional employment opportunities in activities closely interconnected with pineapple cultivation and marketing. For thousands of agricultural workers, the overall employment situation has improved since the 1980s. Apart from generally higher wages, they now get more days of work thanks to additional labor required on rubber plantations intercropped with pineapple. Agricultural workers no longer find it hard to obtain casual work in Vazhakulam. According to my own estimate based on information from farmers and agricultural workers, about 900,000 to

1,200,000 person days are devoted to agricultural work related to pineapple production. This would equal full employment for about 4,000 agricultural workers, but the jobs are dispersed to more people.

Furthermore, agricultural workers have welcomed the shift from paddy to dryland crops (including pineapple) because dryland work is generally less seasonal and better paid than paddy work. Hundreds of poor people are also engaged in activities related to pineapple trading, such as loading and transportation and, to a lesser extent, casual petty trading and brokering. Yet, the access to pineapple-loading jobs is restricted by the labor unions, which have successfully bargained for good wages for the few appointed loaders. Moreover, with the shift to fresh fruit, hundreds of seasonal and permanent jobs in fruit-processing factories in Vazhakulam and elsewhere in Kerala have been lost.

The environmental impact of pineapple cultivation depends on the pineapple-cultivation system and the applied cultivation methods, including the duration of continuous pineapple cultivation. Growing pineapple in contoured rows on upland, and the cultivation of pineapple for a limited period of time, have no severe environmental effects. On former paddies, however, pineapple cultivation has negative environmental consequences, particularly if grown continuously, without crop rotation, over several pineapple life cycles.

Pineapple cultivation on former paddy land necessitates temporary or permanent conversion of wetland into dryland. The severe consequences of these paddy conversions for water, soils and wetland ecosystems, as well as for the flexibility of farmers and the livelihood of other groups, have already been described as a general phenomenon in Kerala (see Section 5.3). In Vazhakulam, the pineapple cultivation on 1,000 or 1,200 acres of former paddies has contributed to these problems. Apart from the general environmental consequences associated with paddy conversions such as water scarcity, increased surface run-off and change of soil properties, the shift to pineapple cultivation involves additional environmental problems. For example, farmers adjacent to pineapple fields mention an increase of pests (e.g., field rats) that use pineapple fields as breeding grounds.

As a result of permanent pineapple cultivation, risks are high that soil nutrients will become depleted. In the case of pineapple, soil depletion is very severe because the restitution of plant material (e.g., leaves) is irregular. Furthermore, land under pineapple cultivation is not covered much with vegetation. This results in changes in soil temperature and moisture, changes that alter the chemical and physical composition of the soil and may also increase soil erosion (Py *et al.* 1984: 314). In Kerala, for instance, upland pineapple monoculture faced declining yields and was discontinued because of soil-fertility problems (my own investigation in Mattanur, Kannur District).

Particularly on former paddies, farmers tend to react to soil depletion and the resulting decline in yields by using chemical fertilizers to excess, but this can destroy the microbiology of the soil. So far, no one knows how long it will take for the soil to recover from a long period of pineapple cultivation. One effective way to mitigate soil-fertility problems on former paddies would be to practice crop rotation. However, only a very few pineapple cultivators rotate their crops. The use of chemical fertilizers also pollutes air and water resources; and its untimely application shortly before the harvest can also produce harmful chemical residues in the fruit. By contrast, the use of the flower-inducing chemical ethephon is not particularly harmful to human health. Ethephon is classified as only slightly toxic for animals and human beings. Moreover, the potential for contamination of groundwater with ethephon appears to be low to moderate (Extoxnet 1993). Moreover, ethephon is applied about four months before the fruit is consumed, so that no residues remain.

Apart from the use of fertilizers and ethephon, increasing application of pesticides and herbicides contributes to the "chemicization" of pineapple cultivation in Vazhakulam. In 1995, about 10 percent of the pineapple growers applied chemical herbicides (e.g., *bromacil, diurion)* in order to save labor costs. Diurion, which is toxic to plants, can stay in soils; bromacil, which is moderately toxic, can contaminate groundwater (Extoxnet 1993). Compared with other pineapple-growing areas around the world, pesticides are rarely used in Vazhakulam. However, cultivators may react with more agro-chemicals because of the spread of pests and plant diseases that is facilitated through the large-scale, pure cultivation of pineapple. For example, the mealy bug *(Pseudococcus brevipes)*, which did not appear in that region until 1995, already in 1996 infected and destroyed about 2 percent of the pineapple fields. The pineapple growers use pesticides only when they notice the pest, which, however, is difficult to recognize. Moreover, the farmers usually do not apply the recommended chemicals but, instead, use inappropriate pesticides that are easily available in the market.

Unlike on former paddies, pineapple does not necessarily cause environmental problems on upland as long as it is cultivated properly (in contoured rows, intercropped). Pineapple intercropping on rubber replantations is unproblematic because it is stopped after four years. Furthermore, pineapple prevents soil erosion if grown in contour lines. However, in some areas of Vazhakulam, pineapple is planted in vertical rows on upland resulting in enhanced soil erosion. Distressing experience with cultivation in vertical rows was noted in the 1930s on big pineapple plantations in Malaysia. There, strong soil erosion occurred and pineapple cultivation eventually had to be given up. Pineapple intercropping on young rubber plantations that was practiced over decades, on the other hand, had no negative environmental effects (Wee 1970, cited in Reichart 1982: 59). Soil erosion on pineapple plantations can also be mitigated through mulching. In important pineapple-producing countries such as Thailand or the Philippines, the bare soil between the pineapple rows is covered with organic mulch or plastic foil.

In sum, the massive growth of pineapple cultivation since the mid-1980s has had positive socioeconomic consequences in Vazhakulam. By contrast, the environmental impact has been ambiguous. The people in Vazhakulam themselves seem to rate socioeconomic concerns higher than environmental ones. They appreciate the massive growth of pineapple cultivation and marketing that has, along with rubber cultivation, brought wealth and employment to the region. The locals trust in the sustained high profitability and marketability of pineapple. Thoughts about environmental consequences are secondary and related to socioeconomic implications that are felt *at present* (i.e., when water scarcity induced by neighboring pineapple fields forces shifts away from paddy cultivation).

Furthermore, farmers in Vazhakulam are not aware of all the environmental problems associated with pineapple cultivation. Long-term effects of continuous pineapple cultivation or of excessive use of chemical fertilizers are only acknowledged by a few pineapple growers. Adverse effects of pineapple cultivation in vertical rows on slopes, on the other hand, are expected; but lessors, leaseholders and agricultural workers with no long-term interest in the land accept the resulting soil erosion and declining land productivity in order to increase their short-term profits. Generally, cultivators do not apply environmentally sustainable cultivation methods, even if these are known, unless they see a close link between the environmental condition and the well-being of their family.

Unsurprisingly, there is no commonly-used Malayalam word equivalent to the term "sustainable development." This is perhaps because people do not think in abstract terms about protecting the environment or maintaining opportunities for future generations, but they do have specific long-term rationales that relate to the well-being of their own children and grandchildren. Farmers therefore consider the maintenance of land productivity to be important, for example, but generally give higher priority to a good education for their children. This may reflect the declining importance of agriculture and of land resources for people's livelihood.

6.4   Relative Influence of the Pineapple Market

This analysis starts with the cultivators' own explanations for why they grow, or do not grow, pineapple, continues with an interpretation of individual motives and relevant cultural values, offers some reflections on socioeconomic factors and, finally, considers technical-material factors. The section as a whole seeks to identify what influence the pineapple market, in relation to other factors, has had on the pineapple boom since the mid-1980s.

6.4.1   The Cultivators' Viewpoint: Reasons for Pineapple Cultivation

Most informants – whether poor, middle-class or rich – spontaneously mention marketability and profitability as the basic reasons for growing pineapple (Mauritius variety). They grow pineapple because it can be sold at any time to nearby traders, and because pineapple production is very remunerative compared with most other crops (except rubber). Relatively quick cash profits can be made with pineapple production. The profitability of pineapple cultivation is further enhanced through the application of new cultivation techniques. Pineapple growers apply ethephon in order to time the harvest to times of expected high demand, as well as to reduce harvesting time and costs. They irrigate pineapple in order to make possible yields during the summer, when seasonal demand and prices are highest.

On rubber replantations, farmers intercrop pineapple (or lease land) because this provides an additional income during a time when the young rubber trees do not yield yet. This intercropping depends on decisions regarding rubber cultivation, which is given priority by the landowners. Furthermore, pineapple is the only cash or food crop for which India's Rubber Board allows intercropping; in the case of other crops, it withdraws the substantial replantation subsidies. In addition, application of fertilizer to, and irrigation of, pineapple has a positive effect on the growth of young rubber trees. This, however, is rather a side-effect than a reason for farmers to intercrop pineapple. Since rubber replantations for pineapple inter-cropping have become scarce in Vazhakulam, cultivators are shifting to other cultivation forms, particularly to pure pineapple cultivation on former paddy land.

On former paddies, the productivity of pineapple cultivation is highest, 10 tonnes per acre as compared with seven tonnes per acre on rubber replantations. But this is only a minor reason for farmers to convert paddies into pineapple fields. Aside from the general profitability and marketability of pineapple, additional reasons for converting paddies are the general non-profitability of paddy cultivation, the severe scarcity and high costs of paddy workers, and the water shortage for paddy cultivation induced by conversion of neighboring paddies into dryland (e.g., pineapple fields). Moreover, a cultural aversion to converting paddies (i.e., the emotional attachment to paddy farming) is slowly fading. One old farmer put it into the following words:

> Formerly, paddy was the pride of us farmers. Even coconut trees were cut down in order to grow (dryland) paddy. That has all changed. Nowadays people are not even ashamed to convert paddies into dryland and to grow pineapple there.

The high investment costs required for the conversion, on the other hand, are still a disincentive for farmers to convert paddies.

On upland other than on rubber replantations, pineapple cultivation is not as popular because productivity is lower here. Regarding intercropping, in particular, tree shade affects pineapple yields. Furthermore, pineapple seems to have a negative side-effect on the growth of coconut trees and other crops. Better productivity of upland pineapple cultivation can only be achieved through costly and labor-intensive inputs such as irrigation or maintenance of terraces. Mainly those cultivators who do not have access to more suitable land (i.e., rubber replantations, converted paddies) grow pineapple on upland.

6.4.2 Interpretation of Individual Motives and the Influence of Cultural Values

Traditionally, pineapple was cultivated in the kayala system in order to check soil erosion. This may reflect a long-term interest in maintaining land productivity. Nowadays, cultivators in Vazhakulam seem to have rather short-term motives regarding commercial pineapple cultivation. The primary reasons given for producing pineapple (i.e., profitability and marketability) seem to indicate that pineapple cultivation is profit-oriented.

In many interviews, cultivators spontaneously applied a profit-maximiza-tion rationality to explain why they cultivate pineapple: "I grow pineapple because it is *most* profitable on this land." Although pineapple is said to be most profitable, no cultivator would fully rely on pineapple cultivation. Less profitable seasonal and annual crops such as banana or tapioca and perennial crops such as coconut or arecanut are still grown in order to diver-sify income and reduce overall farming risks. Cultivators, therefore, seem to be both, risk-reducers (diversifying income, growing low-risk perennial crops) and profit-maximizers (cultivating pineapple on a large scale). Also, in the case of pineapple cultivation, cultivators try to reduce risks of crop failure and low seasonal prices by scheduling two to four harvests per year.

For poor households, the scope for maximizing profits with pineapple cultivation is narrower than for rich or middle-class households. Poor households simply lack the means for large-scale pineapple cultivation. They have limited access to land, capital, labor and irrigation facilities. Access to information and knowledge regarding cultivation methods and prices, on the other hand, seems to be relatively equal between the rich and the poor. Moreover, poor peasants often *cannot afford* to be profit maximi-zers: they must be more risk-averse in order to be able to cope with stress in less fortunate periods. Therefore, poor peasants tend to go for crops that are less seasonal and more continuous than pineapple (e.g., coconut).

Although economic motives in general, and profit-maximization in particular, proved to be the most important motives for growing pineapple, other considerations related to status, prestige and leisure also play a role. One farmer, for example, intercrops banana – and not the more profitable pineapple – on his rubber replantation in order to prove to the Rubber Board that banana intercropping has no negative side effect on the rubber trees. However, by doing so, he loses the substantial replanting subsidies from the Rubber Board. Another respondent leases-in land at a distant place in order to organize his leisure time according to his individual preferences that may contradict social norms: this person, who is known as a teetotaller by his relatives and friends, frequently goes to a toddy shop near his pineapple field, where he can have a drink or two without being seen.

A more common non-economic motive is related to status considera-tions. Earlier, pineapple was considered a "poor people's crop," and middle-class or rich people hardly ever grew it on a large scale. This image of pineapple has changed. Indeed, pineapple has become a symbol of prosperity in Vazhakulam – despite the increasingly low cultural valuation of agriculture in general. Nowadays even wedding-party rooms are some-times decorated with pineapple. This image shift may be traced back to the increased application of new, seemingly "modern," cultivation technology, to capital intensity and to the market orientation of pineapple cultivation. Although it is not the central reason for growing pineapple, the altered crop image may have supported the trend toward pineapple cultivation among middle-class people who otherwise have nothing to do with agriculture.

Kerala's Syrian Christians, who are a dominant community in the midland areas of Kottayam and Ernakulam Districts, are generally known as good farmers who have remained committed to agriculture. This may, to some extent, explain the relatively advanced agricultural development in the midland areas of Kottayam and Ernakulam Districts. However, Hindu farmers seem to be as industrious and committed as Christian farmers. The high level of attention paid to agriculture by all communities in this region may have encouraged the spread there of new, profitable crops, including pineapple.

However, growth of pineapple cultivation in Vazhakulam cannot be explained by cultural values or growing profit orientation. Agriculture in that region had already become highly market- and profit-oriented in the 1950s, when rubber cultivation became important. With the increased use of new hybrid seeds and agro-chemicals in the 1970s, paddy cultivation also became more commercialized. So, the change from a more subsistence-oriented agriculture to a more market-oriented agriculture had already occurred before the pineapple boom started in Vazhakulam. Rather than trying to explain the pineapple boom by the increased profit orientation or motives of individual farmers, or some general changes in community cultural values, I shall therefore focus on specific socioeconomic and technical-material conditions that rendered pineapple cultivation more profitable.

6.4.3 Demand, Relative Prices, Exchange Costs

The cultivators' main reasons for growing pineapple (i.e., profitability and marketability) indicate that pineapple markets are central to the pineapple boom in Vazhakulam. According to the farmers, pineapple has become the second-most profitable crop in that region after rubber. Provided that market conditions are normal (i.e. prices do not drop below Rs. 3.50 per kg), a farmer can achieve annual returns of Rs. 20,000 to Rs. 40,000 per acre for pineapple intercropping on rubber replantations, and Rs. 30,000 to Rs. 60,000 per acre for pure pineapple cultivation on former paddy land (rough estimates based on information from pineapple growers in Vazhakulam). For competing crops such as paddy or tapioca, the average returns may be less than half. However, average returns reflect net profits only partly because cultivation costs are not considered. Generally, cultivation costs are comparatively high in the case of pineapple (see Section 6.4.5).

The profitability of pineapple may be explained by increasing pineapple prices as a consequence of growing demand for fruit in India. Fig. 8 shows the changes in relative prices between pineapple and competing crops since 1974-75.

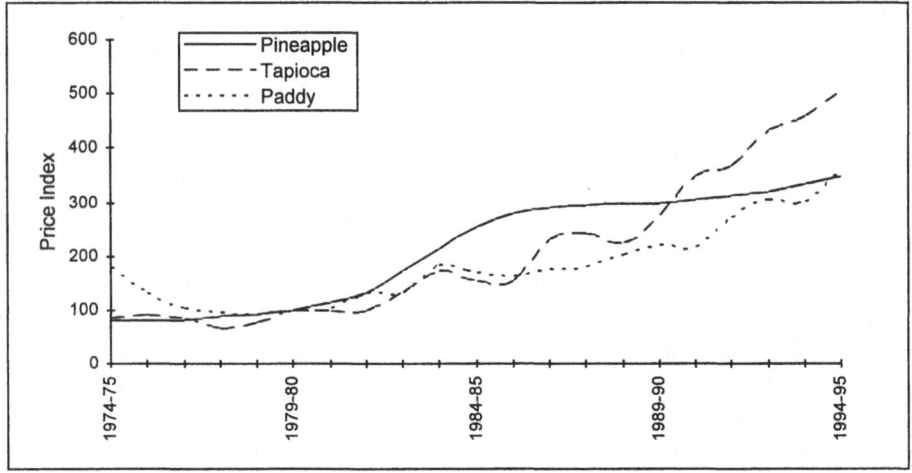

Fig. 8   Prices of Pineapple and Competing Crops
Sources: Pineapple prices: my own estimates based on information from traders and culti-
vators in Vazhakulam. Tapioca and paddy prices: Directorate of Economics and
Statistics (Statistics for Planning, various issues). Index: 1979-80 = 100 (pineapple:
Rs. 1.50 per kg; tapioca: Rs. 0.43 per kg; paddy: Rs. 1.37 per kg).

In the late 1970s, paddy prices decreased in relation to tapioca and pineapple. This may have encouraged farmers to shift away from paddy cultivation. From 1980 to 1986, local pineapple prices increased more rapidly than the prices for competing crops (paddy, tapioca). This may have motivated cultivators in Vazhakulam to grow more pineapple. In other periods, however, changes in relative prices between pineapple and other crops were less significant. Although pineapple prices have increased only slowly since 1986 (particularly in relation to tapioca prices), pineapple cultivation has continued to expand whereas tapioca and paddy cultivation have not. This implies that cultivators did not immediately react to changes in relative prices and/or that other factors than changing relative prices were also important for the pineapple boom in Vazhakulam.

However, the accuracy of the historical data on pineapple prices is doubtful because it has to rely on people's memory. Furthermore, average local pineapple prices in Vazhakulam may have been different from those in the rest of Kerala. Fig. 8 does also not reflect seasonal and daily price fluctuations, which are comparatively high and often unpredictable in the case of pineapple. Prices are mainly dependent on the pineapple terminal market in Mumbai. Depending on demand and supply, the "normal" price fluctuation were between Rs. 3.50 and Rs. 6 per kg in 1995. In the recent past, however, prices in Vazhakulam occasionally dropped suddenly to Rs. 1.50 per kg during crises such as the communal riots in Mumbai (December 1992, January 1993), the plague in North India (October 1994), or national truck strikes, the last of which took place in April 1997, when truck drivers agitated against a proposed national service tax on road transport. This shows that price fluctuations can also be dependent on non-economic factors. Although pineapple growers apply risk-spreading practices such as distributing pineapple yields over the year, such institutional weaknesses can leave them in dire straits from time to time. However, it seems that for middle-class and rich farmers, the comparatively high average returns of pineapple cultivation offset the high risks due to fluctuations in pineapple prices.

Apart from the mentioned risks, transaction costs for pineapple market-ing have generally become lower in India, including Kerala. Improvements in transportation were critical for this highly perishable commodity. The general development of the national highway network, the growth of inter-state trade, and the start of flower exports from Tamil Nadu via the international airport in Thiruvananthapuram since the early 1980s have improved transportation by increasing the availability of empty trucks that can take goods from Kerala on their return journey. Earlier, pineapple had to be transported by railway for sale across the state borders. Transportation of goods by the state-owned Indian railways requires more formalities and tends to be less reliable than transportation by private road-transport companies. Thanks to the increased availability of road transport since the 1980s, therefore, exchange costs for the inter-state pineapple trade have declined significantly.

Improved transportation may have contributed to better availability of fresh-pineapple fruit on urban markets. Eventually, the shift by consumers from canned pineapple to the cheaper fresh pineapple contributed to the decline of pineapple processing in Kerala and in India generally. Furthermore, the change from canned pineapple to fresh pineapple involved changed demand from Kew to Mauritius pineapple. To deal with altered consumer preferences, growers of Mauritius pineapple in Vazhakulam increasingly used ethephon and irrigation to adjust production to the seasons of highest demand – the hot summer months from March to May and the times of religious and cultural festivals – rather than allowing the bulk of their product to come to market during the natural harvest season which lasts from May to July (see Fig. 9).

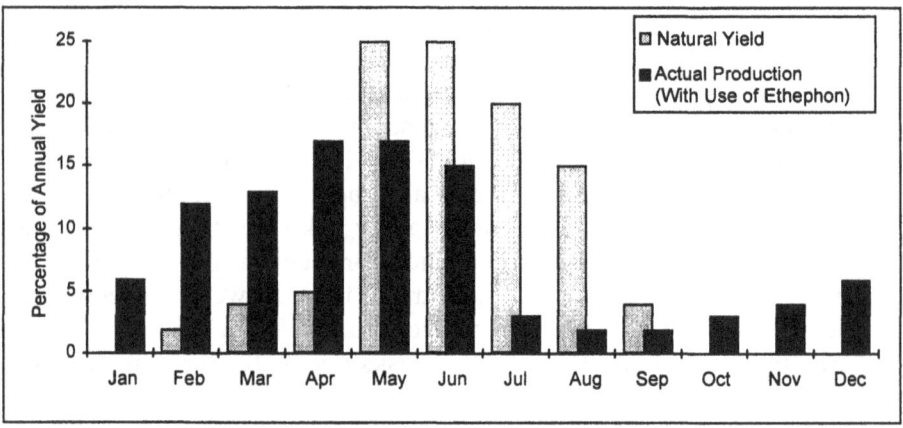

Fig. 9   Monthly Mauritius-Pineapple Production
Source: Estimates based on information from local traders and cultivators.

The new market conditions since the mid-1980s set the pace for pineapple production in Vazhakulam. However, although these new market conditions applied throughout Kerala, pineapple cultivation picked up substantially only in Vazhakulam. Indeed, in most other parts of Kerala, the market-ability of pineapple has remained bad or has even worsened. In other words, pineapple markets have not emerged "automatically" as a response to new market conditions. To explain this, it is necessary to look in greater detail at the development of the "real" pineapple market in Vazhakulam and at verti-cal relations in the market system.

6.4.4 The Making of a Pineapple Market

In Vazhakulam, experience with pineapple marketing since the 1960s and 1970s (see Section 6.2) considerably eased the growth of pineapple marketing in the mid-1980s. Experienced traders were able to take advantage of the new market conditions, which turned pineapple marketing into a business that depends on daily information about demand and requires intensive contacts with fresh-fruit wholesale agents in terminal markets. In particular, the local pineapple merchants have worked hard to establish direct and personal connections with fresh-fruit wholesalers, especially in terminal markets outside Kerala. The high quality and special traits of Mauritius pineapple, which is not grown commercially elsewhere, helps them. Today, most of the 200 or 300 tonnes of pineapple traded daily in Vazhakulam go through the hands of 50 large wholesalers, who have specialized contacts to one or two of the various terminal markets. The big wholesale merchants buy either directly from the thousands of cultivators or from the hundreds of petty traders (e.g., country buyers, commission agents), most of whom are only temporarily engaged in the pineapple trade.

In Kannur District, by contrast, where Kew pineapple was grown as an intercrop on extensive cashew plantations until the mid-1980s, the experienced local pineapple traders were not able to make use of the new market conditions. The reason is that farmers were reluctant to shift to Mauritius pineapple, whose spiny leaves make the collection of nuts from the ground of cashew plantations very hard. Rather than continuing to grow the more suitable, smooth-leafed Kew variety, most farmers stopped cultivating pineapple altogether. Soon, local pineapple production was so small that profitable marketing became impossible. As a result, both pineapple cultivation and pineapple marketing collapsed in Kannur District.

Repeated dealings, more personal contacts, and the development of telecommunication have all lowered transactions costs in the growing trade in pineapple between Vazhakulam and the major urban terminal markets. Also needed to make large-scale pineapple trade possible has been the informal organization of Vazhakulam's wholesale merchants into seven business groups. These groups allow the risks, which would be too high for any individual, to be spread over a larger number of traders. Each partner of the business group specializes and develops personal contact in a different terminal market. By pooling information and contacts, the partners can exchange pineapple among themselves in line with demand and prices in all the terminal markets. In other words, each cooperating merchant is able to maintain personal trade links with a particular terminal market and, at the same time, has access to a diverse number of markets, thus spreading the risks.

Special arrangements have also evolved between producers and traders. Two basic types of pineapple-marketing arrangements are prevalent. In the case of big harvests of more than two tonnes, pineapple traders (mostly, the big wholesale merchants) buy directly from the fields, some of which may be more than 50 kilometers away from the marketplace in Vazhakulam. The pineapple grower or a broker informs the merchants about fields that are ready to be harvested. Then, by telephone or during a field visit of the merchant, a price per kg is fixed and the harvest can begin immediately. Harvesting labor is usually arranged for and paid by the farmer, loading and transportation by the trader. Normally, the farmer gets an advance payment (25-35 percent). The amount due is paid one or two weeks after the harvest, that is, when the merchant from Vazhakulam gets the payment from the wholesaler in the terminal market. Long-term arrangements between pineapple growers and traders are exceptional. In these few cases, a firm price per kg is fixed before the application of ethephon.

In the case of relatively small quantities up to two tonnes, the cultivator, himself, has to bring the product to the trader by auto-rikshaw or by jeep. The price is the same as in the case of big harvests, but the cultivator has to bear additional costs for harvesting, loading, transportation to the marketplace, and unloading. Cultivators check the prices at three or four shops in Vazhakulam and in nearby pockets before they sell. In this way of marketing, the pineapple farmers get immediate payment.

In addition to the marketing arrangements between cultivators and traders, the organization of loading and transportation is an important component of pineapple marketing. Pineapple trade has to be carried out very quickly because the pineapple fruit is very perishable and because refrigerated storage rooms are not available. Usually trade-union workers are employed by the trader to grade the pineapple fruits according to their size and to load the fruits on the truck. Specialized transportation companies, which have contractors and booking offices in Vazhakulam, organize the transportation by truck.

These local institutions that evolved to organize and coordinate harvesting, loading and transport of pineapple have had a positive effect on the marketability of pineapple. As a consequence of conflicts, formal local organizations representing the pineapple growers, the merchants and the loaders have been established recently. At the time of my field study, however, no conflicts between the local lobby groups erupted, and pineapple marketing went on smoothly. Now, these local organizations can be regarded as instruments to mediate among the different interest groups and to avoid conflicts that might disrupt pineapple marketing.

From the cultivators' viewpoint, pineapple marketing is benign and efficient. In Vazhakulam, cultivators have never had a problem in selling Mauritius pineapple. Moreover, competition among traders, and low transaction costs due to the well-established pineapple-marketing system, mean that prices in Vazhakulam tend to be higher than elsewhere in Kerala. Although merchants within the business group decide on common farm-gate prices, no monopsonistic situation has emerged. This is because there is competition among several business groups.

Moreover, pineapple growers are not totally dependent on certain pineapple traders, whose role is usually limited to marketing and does not involve moneylending, land-leasing, etc. In other words, the pineapple market is not interlocked with the credit, land or labor markets, and the commercialization of pineapple cultivation does not involve the compulsive involvement in markets. Farmers are therefore free to engage in, and disengage from, the pineapple market and to sell to whoever offers the best terms. "Best terms" means "the highest price" only in the case of small harvests of less than two tonnes and spot transactions. In the case of bigger harvests, which involve later payment, pineapple producers deal only with affluent merchants whom they know well and whom they trust. Repeated dealings and personal relations are also important between traders and farmers. The spatial closeness between trader and farmers encourages personal relations and eases the exchange of information as well as the supervision and enforcement of contracts.

The predominance of Vazhakulam as a regional pineapple-marketing center may be reinforced by the ongoing growth of pineapple production in adjacent areas and by the fact that traders in new pineapple-growing areas may find it difficult to bypass the powerful big pineapple-wholesale merchants in Vazhakulam. These merchants protect their crucial connections to fresh-fruit agents in terminal markets and keep their business secrets. This is easier than protecting the highly visible and reproducible pineapple-cultivation practices of the leaseholders of Vazhakulam. Yet, in the new pineapple-growing areas, local agents, small traders and brokers may go into partnership with the big merchants in Vazhakulam.

Despite the power of the big merchants in Vazhakulam, pineapple marketing may change. The Kerala Horticultural Development Programme (KHDP), for example, plans to establish an alternative marketing channel that would involve the purchase of pineapple, at an annually fixed price, directly from farmers with landholdings of more than 0.5 acres. This would neutralize daily price fluctuations, which are the main marketing problem for pineapple growers in Vazhakulam. However, the KHDP is attempting to motivate the pineapple cultivators to sell their produce only to their own fruit-processing factory, which is due to start in November 1997. In this way, the cultivators' flexibility to choose among many buyers may be reduced. While pineapple growers showed interest in this new marketing system in 1994, when open-market prices were low, they seem to have become indifferent once prices for pineapple increased again after 1995.

6.4.5 Production Costs, Labor Scarcity and Land Relations

As already indicated, profitability does not depend only on returns, but also on production costs. Table 4 summarizes the costs of pineapple cultivation. These estimates are intended to give a general impression of the capital intensity of pineapple production, but they should not be mistaken for a precise calculation. Moreover, production costs vary significantly among pineapple cultivators.

Table 4   Annual Pineapple-Cultivation Costs (Intercropping)

Type of Costs	Rs. per Acre (range)		
Labor	12,000	-	16,000
Land Lease	2,000	-	6,000
Planting Material*	2,000	-	3,500
Chemical Fertilizers	0	-	3,000
Irrigation	0	-	3,000
Flower-Inducing Chemical	100	-	150
Total Costs	16,100	-	31,650
Average Returns	*20,000*	-	*40,000*

Source: *Estimates based on information of pineapple cultivators in Vazhakulam. (*Actual planting-material costs of Rs. 6,000 to Rs. 10,500 per acre projected over three years.)*

Table 4 shows that total annual costs for pineapple intercropping on rubber replantations are between Rs. 16,000 and Rs. 32,000 per acre while average returns are between Rs. 20,000 and Rs. 40,000 (see Section 6.4.3).

However, in the initial year of pineapple cultivation, when the land has to be prepared and planting material must sometimes be purchased, cultivation costs may be higher. Furthermore, cultivation costs for pineapple on former paddy fields tend to be higher than on rubber replantations because of more expensive leases and increased fertilizer use. On the other hand, minimum costs may be significantly lower than reflected in Table 4 if the planting-material does not have to be purchased (but can be taken from old fields), the cultivation is on land owned by the cultivator (not leased-in land), much family labor is used (not only hired labor), and herbicides are employed to reduce labor costs.

Compared with other crops, the cultivation costs for pineapple are very high. Local informants maintained that cultivation costs for paddy or tapioca may be only about half of the pineapple-cultivation costs. These high costs prevent poor peasants from engaging in large-scale pineapple cultivation. Moreover, poor peasants usually do not qualify for formal bank loans that require collateral.

As for many other crops, labor costs are very high compared with other factor costs. Moreover, as a consequence of the bargaining strength of the agricultural workers and their unions, in Kerala, labor costs have generally increased more rapidly than agricultural-commodity prices and than most other factor costs. However, trade unions in Vazhakulam do not need to struggle harshly for higher wages in the case of pineapple work. Pineapple growers usually grant pay increases without much opposition. In 1995, daily wages for six hours labor on pineapple fields varied between Rs. 50 to Rs. 70 for women and Rs. 70 to Rs. 90 for men. Sometimes a meal is also provided. The bulk of the manual work on the pineapple fields is carried out by wage laborers, who are hired on a daily and sometimes on a weekly basis. Many of these agricultural workers are very skilled in cultivation techniques. Some of them are even more knowledgeable concerning pine-apple-cultivation methods than the supervising farmers (e.g., leaseholders who generally have little to do with agriculture). Particularly for poor and lower-middle-class pineapple growers, however, family labor is an important supplement to hired wage labor.

Because of increasing labor costs and labor scarcity, farmers in Kerala have generally shifted to labor-saving crops. The growth of pineapple cultivation in Vazhakulam contradicts this general pattern. Pineapple cultivation is relatively labor-intensive, requiring 150 to 200 person days of hired workers per acre annually (my own estimate) compared with 176 for paddy, 140 for tapioca and 82 for coconut (Directorate of Economics and Statistics 1992b). Yet, these average figures obscure seasonal demands and the special requirements related to particular crops. While labor requirements for pineapple cultivation are spread relatively equally throughout the year, for example, paddy cultivation demands much labor during short peak periods and therefore requires a bigger reservoir of agricultural workers. Moreover, pineapple cultivation, unlike paddy cultivation, does not depend on the prompt completion of certain operations, and so does not require that workers be available on a round-the-clock basis. Generally, the agrarian relations of the post-land reform era in Kerala are more suitable for pineapple than for paddy cultivation. In the words of certain agricultural workers themselves, pineapple cultivation is more "contemporary" than paddy cultivation, which they still associate with semi-feudal production relations. Wage laborers also prefer the work in pineapple fields because it is less seasonal and less arduous. These factors explain why farmers in Vazhakulam may face labor scarcity for paddy cultivation while labor for pineapple cultivation is available.

Considering the population density of about 600 people per square kilometer, land scarcity is another issue in Vazhakulam. As elsewhere in rurban Kerala, land-poor people interested in agriculture find it difficult to acquire land. In Vazhakulam, land prices have been forced up to about one to two million rupees per acre by land speculation and the influx of Gulf money. However, medium-term land-leasing arrangements have evolved (despite the legal ban on land tenancy) for pineapple cultivation. These leasing arrangements seem to be beneficial for both lessors (who can receive risk- and trouble-free additional income from land while keeping it as an asset) and leaseholders (who get access to affordable land resources).

According to an estimate by the KHDP, about 30 percent of pineapple production in Vazhakulam is done by leaseholders. These people are often middle-class cultivators, such as salaried employees, with little land of their own. Leasing has evolved particularly for this crop because pineapple cultivation is very profitable and therefore enables lessors to demand a relatively high rent and to achieve adequate returns from the land (the annual rent for land on rubber replantations varies between Rs. 2,000 and Rs. 6,000 per acre; for former paddy land, it may be up to Rs. 8,000 due to the higher productivity there). Moreover, lessors of rubber replantations, where leasing for pineapple cultivation is most widespread, do not have to fear losing land in legal battles with former leaseholders, who may refer to the invalidity of tenancies under Kerala's Land Reform Act of 1970. In any event, lease arrangements, unlike for other crops, are naturally term-limited for pineapple intercropped on rubber replantations: intercropping must be discontinued for ecological reasons after four years, at which time the lessor's rubber trees are fully grown and serve as an unmistakable mark of landownership. Land for pineapple cultivation is usually leased for the whole period of four years, the productive life span of pineapple plants. Poor households usually lack the capital to pay the rent in advance. A very few of them engage in sharecropping contracts, which seem to be limited to pineapple cultivation on upland other than rubber replantations.

Furthermore, lessors find it easy to supervise the cultivation practices of pineapple leaseholders. Especially rubber-plantations owners keep a sharp eye on the leaseholders and insist on pineapple-cultivation methods, such as planting in contoured rows, that have positive effects on the land's productivity and the growth of the young rubber trees. In exceptional cases, however, landowners show no interest in long-term land productivity – an attitude that may be enhanced by uncertainties regarding property rights. The following case illustrates how vaguely defined property rights induce short-sighted cultivation methods that aim at maximizing profits in the near term but endanger sustainability.

> The ownership of 100 acres of land on a hill in Vazhakulam has been under dispute for about 50 years. In the 1950s, a former tenant took his claim on ownership rights to court. However, court decisions were delayed. Due to the ongoing uncertainty, both parties left most of the disputed land fallow, and on the small portion of cultivated land, neither of them grew perennial tree crops, which would be economically and ecologically suitable there. In 1985, finally, a court decision brought clarity: 40 acres remained in the hands of the former landlord; the other 60 acres went to the former tenant, whose family members, however, are disputing about this land now.
>
> Nowadays, most of the 100 acres are leased to various pineapple growers. On the 40 acres where property rights have been clear since 1985, the landowner compels the leaseholders to plant pineapple in contoured rows on rubber replantations. On the 60 acres that are still disputed, on the other hand, the leaseholders are free to grow pineapple as they wish. Eventually, pineapple is grown in vertical lines – a practice that reduces labor requirements since no contour bunds need to be built and maintained, weeding is easier from top to bottom than on a flat platform, etc. This cultivation method, however, enhances soil erosion and reduces land productivity in the long run – a fact that lessors, leaseholders and agricultural workers are very well aware of, but nevertheless allow to happen in order to reap higher immediate profits.

Compared with growing pineapple on one's own land, leasing is less profitable on average, but leasing has the advantage for the landowner of reducing cultivation costs, saving time and avoiding bother and risk. Furthermore, lack of skills regarding new pineapple-cultivation methods (as well as lack of knowledge about possible returns from pineapple cultivation) discourages landowners to engage directly in pineapple cultivation. However, many landowners imitating the cultivation techniques of their former leaseholders, are now starting to grow pineapple on their own. Therefore, the availability of land for lease will probably decline – not only in Vazhakulam but also in adjacent areas, where leaseholders from Vazhakulam have introduced pineapple cultivation. In 1996, two people from Vazhakulam even started to lease-in land outside Kerala (i.e., in Tamil Nadu and Maharashtra. In this way, commercial cultivation of Mauritius pineapple is spreading.

To summarize, pineapple cultivation fits in comparatively well with existing agrarian relations in Kerala. Various adaptations in land- and labor relations have supported the growth of commercial pineapple cultivation in Vazhakulam. The "pineapple boom" calls in question some simplistic generalizations about Kerala: that its farmers are shifting to labor-saving crops, or even, that Keralites have lost all interest in agriculture.

6.4.6 Technology Development and Extension

In addition to the socioeconomic factors discussed above, production technology has also led to changes in the profitability and marketability of pineapple. It is remarkable that, in the case of pineapple cultivation in Vazhakulam, the cultivators themselves, rather than public-sector agricultural researchers and extension workers, have been the main agents in the development and spread of new techniques.

In 1967, the state of Kerala began its first interventions directed at pineapple cultivation. The Development Blocks introduced Kew pineapple into many areas of Kerala, where Mauritius pineapple had been more widespread. Until the early 1980s, state agencies (after 1972, Kerala's Department of Agriculture) disseminated pineapple-planting material free of cost. Their aim was to popularize the cultivation of Kew pineapple in order to support the developing fruit-processing industry in Kerala. In 1980, the Kerala Agricultural University (KAU) introduced the flower-inducing chemical ethephon and formulated recommendations regarding "scientific" pineapple-cultivation methods based on tests with Kew pineapple. These recommendations about planting density, fertilizer use, application of flower-inducing chemicals, etc., have not been adapted yet to the cultivation of Mauritius pineapple, which has become more popular in the meantime. For instance, the KAU recommendations of 1980 were still published in the latest edition of *Package of Practices Recommendations: Crops-1993*, which is the official reference book for Kerala's agricultural extension. In Vazhakulam as elsewhere in Kerala, however, the agricultural officers and demonstrators had other priorities than advocating better pineapple-cultivation methods. Moreover, pineapple growers in Vazhakulam have meanwhile become more knowledgeable about pineapple cultivation than the extension authorities. Only in 1995, the Pineapple Research Station in Vazhakulam, a branch of the KAU, started to develop new recommendations for pineapple cultivation that are based on on-farm trials with Mauritius pineapple.

Nevertheless, the recommendations of the KAU have been influential, though in an unexpected way. Farmers of Vazhakulam made their own field experiments with Mauritius pineapple and considerably modified the suggestions of the KAU. The most important modification relates to the purpose of ethephon application. Pineapple cultivators use ethephon to time pineapple harvests *to expected peak demand* – and not only to have all pineapple plants in a field ready for harvest at once. Furthermore, farmers plant pineapple less densely, but apply more than twice as much chemical fertilizer as recommended. The spread of these modified cultivation techniques takes place from farmer to farmer, without the intervention of public agricultural extension officers. New pineapple growers get information from more experienced cultivators and agricultural workers and imitate their cultivation methods. In new pineapple-growing areas, landowners learn about the new pineapple-cultivation methods developed in Vazhakulam through the leaseholders from there. Consequently, Vazhakulam farmers are no longer able to keep the new pineapple-cultivation techniques secret among themselves.

The example of pineapple cultivation puts the role of state agencies for developing and extending agricultural technologies into perspective by stressing the indigenous farmers' potential to adapt and modify "imported" technologies. Generally, technological innovations have increased the productivity, and thus also the profitability, of pineapple cultivation, while the possibility of producing pineapple throughout the year, and timing harvests to peak demands has enhanced the marketability of this fruit. Growth of pineapple production and improvement of marketing opportunities have mutually reinforced each other.

6.4.7 Diminishing Importance of Bio-Physical Conditions

The climate in Vazhakulam has proved to be suitable for the cultivation of pineapple despite relatively high annual rainfall. Thanks to proper drainage on upland and on former paddies, the Mauritius pineapple has flourished there. Furthermore, the lateritic soils, which are widespread in Vazhakulam, are optimal for pineapple cultivation. It is said that the very particular soil characteristics in parts of the upland of Vazhakulam are exceptionally suitable for the production of pineapple with good taste and high sugar content. Earlier, the high quality of pineapple from Vazhakulam motivated agents of processing factories to buy local pineapple. However, the widespread use of chemical fertilizer has meant that fruit quality is now equalized and less dependent on soil characteristics. Indeed, pineapple cultivation has recently spread more rapidly in paddies than in the naturally more suitable lateritic soils. The application of new cultivation techniques would make pineapple cultivation just as productive elsewhere in Kerala as it is in Vazhakulam. Nevertheless, pineapple from Vazhakulam still has a good reputation among consumers.

In sum, therefore, only the initial moderate growth of pineapple cultivation was influenced by favorable bio-physical conditions (i.e., the exceptional soil qualities). The massive growth pineapple cultivation since the mid-1980s, on the other hand, has been the result of socioeconomic and technical factors as well as of the proactive role played by traders and cultivators in Vazhakulam. The most important causes for the pineapple boom have been the new market conditions (i.e., growing demand for fresh-fruit pineapple, favorable relative prices for pineapple, reduced exchange costs), the active engagement of traders in building up large-scale pineapple marketing from Vazhakulam, and technological innovations and modification by pineapple growers. Besides, pineapple cultivation has been comparatively suitable under the existing agrarian relations in Kerala. Conditions such as the Panchayat pineapple marketplace (built in 1982), regulations of the Rubber Board allowing pineapple intercropping on rubber replantations (1985) and the general shift away from paddy (since the 1970s) have also facilitated the pineapple boom in Vazhakulam. So, today's favorable conditions for pineapple cultivation there are *man-made* rather than *natural*. It was these conditions that made it possible for pineapple cultivation to become a common agricultural practice among farmers in Vazhakulam.

6.5 Sustainability, Markets and Pineapple Cultivation

Generally, the partly market-induced, massive growth of pineapple cultivation has had positive socioeconomic consequences for people in Vazhakulam, creating new, remunerative business and employment opportunities. The pineapple market did not discriminate directly against poor peasants, who, however, find it difficult to start commercial pineapple cultivation because of its high capital intensity.

At the same time, however, the development of Mauritius-pineapple cultivation and marketing in Vazhakulam resulted in comparative disadvantage for people in previous pineapple-growing, -marketing and -processing regions. The growth of Mauritius-pineapple cultivation was initially concentrated in Vazhakulam. However, the people of this region now find themselves increasingly in competition with people outside Vazhakulam who are beginning to grow pineapple on a commercial scale in areas where the growth potential is still high.

It is hard to predict the future of pineapple cultivation and marketing. The pineapple business is unlikely to remain so profitable in Vazhakulam, once new supply areas in and outside Kerala develop. Furthermore, long-term trends regarding consumer demand are unknown. Much also depends on the ability of merchants to get access to new terminal markets, including export markets: the planned Nedumbassery International Airport will be only about 30 kilometers away.

Generally, the increased dependence on the pineapple business has made the people of Vazhakulam vulnerable to processes over which they have little control. An abrupt decline or long-term downward trend in pineapple marketing would adversely affect the livelihood of many people in Vazhakulam, including cultivators, agricultural workers, traders and loading workers. However, cultivators may be able to shift back from pineapple to other crops quickly, if pineapple becomes less profitable and less marketable in the long run.

The environmental impact of pineapple cultivation depends on the cultivation methods applied. Pineapple cultivation in contoured rows on upland (including rubber plantations), as well as the cultivation of pineapple for a limited period of time, have no adverse environmental effects. Permanent pineapple cultivation on former paddies that is often accompanied with excessive fertilizer use, however, has negative environmental consequences, including depletion of soil nutrients, soil and water pollution with agrochemicals, abandonment of wetland ecosystems leading to water scarcity as well as increased likelihood of floods downstream, and reduced flexibility for cultivators at present or in future.

The pineapple market helped to produce the massive growth of pineapple cultivation in Vazhakulam after the mid-1980s, and thus contributed indirectly to some negative environmental processes that have socioeconomic implications. Moreover, increased pineapple trade requires motorized transportation, which leads to well-known environmental problems like fuel depletion and air pollution. On the other hand, improved transportation facilities and increased mobility have positive social "externalities."

The specific environmental problems associated with pineapple cultivation derive from particular agricultural practices (i.e., paddy conversion in order to grow pineapple, continuous cultivation of pineapple, excessive use of chemical fertilizers, planting in vertical rows). However, these unsustainable cultivation practices, which affect the flexibility of paddy farmers and the development opportunities of future generations, were not specifically encouraged by the pineapple market, neither by its structural properties nor by special demands of traders or consumers. Rather, other factors (e.g., a lack of awareness of, and knowledge about, the negative effects of excessive use of agro-chemicals and about the environmentally positive effect of crop rotation, new agrarian relations inducing the shift away from paddy cultivation, undefined property rights) played a direct role in affecting the environment. Nevertheless, the pineapple-assembly market in Vazhakulam did have a regionally concentrating effect on pineapple cultivation. It encouraged the shift toward pineapple cultivation on former paddies, once pineapple intercropping on rubber replantation in Vazhakulam could no longer expand.

The illustrated case of commercialized pineapple cultivation in Vazhakulam may challenge some hypotheses of the various viewpoints that I have discussed in Section 3.6. The findings are inconsistent with the Marxist-influenced viewpoint suggesting that environmental problems are caused by poverty or unequal access to resources. Rather, it is the middle-class and rich cultivators who grow pineapple on a commercial scale, both in sustainable and unsustainable ways. As a consequence of commercial pineapple cultivation, poor peasants were neither displaced nor did they come under the "simple reproduction squeeze" that would result in overuse of natural resources. The pineapple market is not a compulsion but rather an opportunity, although, admittedly, one that poor peasants are not able to make full use of because cultivation costs for pineapple are very high.

The neoclassical-economist viewpoint has emphasized that under-pricing agricultural inputs, natural resources and environmental goods leads to the adoption of unsustainable practices such as the excessive use of chemical fertilizers. Increasing fertilizer prices as a result of India's cuts in subsidies after 1991, however, did not motivate all cultivators in Vazhakulam to reduce the fertilizer dosage, because excessive fertilizer application to pineapple still pays off – at least in the short term. Only cost-averse pineapple growers reduced the use of chemical fertilizers. As a consequence, however, some pineapple fields receive too little fertilizer, resulting in soil-nutrient depletion and declining yields. In the case of fertilizers, therefore, price signals seems to be an unreliable instrument to achieve "optimal" fertilizer use, which depends largely on the soil type and on the crop grown.

Moreover, the neoclassical-economist viewpoint underestimated the significance of unacknowledged conditions and unintended consequences for farmers' decision-making. Because they lacked long experience with large-scale pineapple cultivation, many farmers adopted cultivation methods that have unintended and unacknowledged consequences and may adversely affect yields on their own land. These unsustainable practices cannot be interpreted as deliberate actions to "externalize" environmental costs. This example implies that knowledge and information about the long-term consequences of particular agricultural practices may be as important as price incentives or clearly defined property rights for sustainable development.

The spread of commercial pineapple cultivation may also be interpreted as the substitution of unsuitable "modern" cultivation technology for sustainable traditional practices such as the kayala system. Yet, pineapple intercropping on rubber replantations can be regarded as a further development of the kayala system. Moreover, the indigenous modification of particular pineapple-cultivation methods (i.e., pineapple cultivation on former paddies, higher fertilizer use than recommended) has had a negative effect on the environment and subsequently on the flexibility of farmers and future generations. In these cases, "outside" agencies such as the Krishi Bhavan or the Pineapple Research Station seem more aware of, and concerned about, environmental and long-term, socioeconomic effects than the farmers. One cannot, therefore, assume that farming practices based on indigenous knowledge will invariably be eco-friendly or particularly adapted to local bio-physical conditions.

7 Cashew Cultivation in Mattanur-Iritty

7.1 General Patterns of Cashew Production and Trade

Cashew is cultivated in about 25 countries in South and Southeast Asia, East and West Africa, and South and Central America. Large-scale processing and trading of cashew began in South India in the 1920s. Since then, global production and trade of raw cashewnut and of processed cashew kernel have grown rapidly. From 1980 to 1995, for example, global raw-cashewnut production increased from about 450,000 tonnes to almost 700,000 tonnes (FAO 1990-1996). The most important cashew-growing countries are Brazil and India. In 1994, almost 30 percent of all raw cashewnut produced in the world was exported for processing. The main importer of raw cashewnut is India; the main exporters are Vietnam, Indonesia, Tanzania and Guinea Bissau. India also boasts by far the biggest cashew-processing industry.

In India, the cashew tree, which was introduced in the 16th century by Portuguese colonialists, spread without much human propagation before commercial cashew processing began in the 1920s. South Kerala. From its very beginnings, the cashew industry has been export-oriented. In the glory days of the mid-1960s, cashew-processing factories in the town of Kollam (South Kerala) accounted for nearly 90 percent of the global cashew-kernel trade (Kannan 1981: 109). Since then, the cashew industry there has suffered the effects of competition from processing factories in East Africa, in Brazil and, more recently, in Vietnam. Cashew-processing factories in Kerala also have shifted to neighboring states (Tamil Nadu, Karnataka and Andhra Pradesh). In 1977, affected by new limits on the supply of raw nuts from East Africa, India's cashew exports and production sharply declined and remained low until 1988. Since 1989, aided by new supplies from South Asian and West African countries, the Indian cashew industry has recovered – and has recently grown very much. India remains the world's largest producer of cashew kernels, holding about 65 percent of the global market. In 1994-95, about 77,000 tonnes of cashew kernels worth Rs. 12.4

billion (US\$ 400 million) were exported. Cashew exports accounted for 1.5 percent of India's total export earnings. In 1994-95, cashew was the biggest foreign-exchange earner of all agricultural products from India, surpassing even rice and tea (Cashew Bulletin 1995, no. 8: 1).

However, at the same time, the Indian cashew industry is a significant global importer of raw cashewnut, because domestic raw-cashewnut production is not sufficient to meet the total demand of industry. In 1994-95, about half of the raw nuts were imported from East and West Africa and from Southeast Asia. Raw-nut imports amounted to about Rs. 6.9 billion, reducing India's net cashew-export earnings to Rs. 5.5 billion (Cashew Bulletin 1996, no. 2: 2). Since the "crisis" of the cashew industry between 1977 and 1988, the high dependence on raw-cashewnut imports has been the main problem of India's cashew industry. Supplies are not guaranteed because new raw-cashewnut producers such Vietnam are establishing and protecting their own cashew-processing industry.

Already in the mid-1960s, India's government launched a cashew-development program in order to reduce the high dependence of the Indian cashew industry on imported raw nut. In 1966, the central government established the Directorate of Cashewnut Development, a subdivision of the Union Ministry of Agriculture. The duty of the Directorate, which has its headquarters in Kochi, is to implement the Centrally Sponsored Cashew Development Programme that aims at enlarging the area under cashewnut cultivation as well as at improving the productivity of cashewnut cultivation. Subsequently, large new plantations were established in former waste lands in Maharashtra, Orissa and Andhra Pradesh. Since 1981, furthermore, over 25 high-yielding varieties have been developed by India's Agricultural Universities and Cashew Research Stations (Indian Cashew Journal 1994, no. 3: 12).

According to India's Directorate of Cashewnut Development, between 1966 and 1992, the area under cashew in India was more than doubled. In the same period, raw-cashewnut production increased threefold. In 1966, Kerala produced about 77,000 tonnes (70 percent of the total production in India). Until today, Kerala has remained the most important cashew-growing state, producing about 152,000 tonnes (43 percent of the national total) despite the newly established cashew plantations in central India under the Cashew Development Programme. The Directorate of Cashewnut Development has estimated that the area under cashew cultivation in Kerala is about 390,000 acres.

According to Kerala's Directorate of Economics and Statistics, however, the area under cashew cultivation in Kerala is only 275,000 acres, and the annual production of raw cashewnut is only 100,000 tonnes. Moreover, the Directorate of Economics and Statistics has estimated that the area under cashew cultivation and production of raw cashewnut in Kerala has not increased significantly in the past 30 years. Such discrepancies in the estimates make it impossible to assess the development of cashew cultivation in Kerala as a whole. This study therefore confines itself to the appraisal of trends at the local level.

In Kerala, cashew is grown in midland and lowland regions. The main cultivation areas are in the northern districts, namely in Kannur and Kasaragod Districts. In South Kerala, many cashew plantations have been replaced with rubber plantations since the 1950s. The productivity of cashew in Kannur District is among the highest in the world: about 590 kg per acre compared with an average of 351 kg per acre in Kerala and 260 kg per acre in India (Directorate of Economics and Statistics; Directorate of Cashewnut Development). In some cashew-growing regions of the world, productivity is as little as 40 kg per acre (Ohler 1979: 91). North Kerala's comparatively high cashew yields are attributable to favorable bio-physical conditions rather than to the application of intensive cultivation methods. Potential maximum yields might be as high as 1,200 to 1,600 kg per acre (Salam & Mohanakumaran 1996). Raw cashewnut from Kannur District is also of a very good quality, probably among the best in the world.

In Kerala, raw-cashewnut marketing was under state control from 1976 to 1995. In 1976, the State Government of Kerala declared raw cashewnut an "essential article" under the Kerala Essential Articles Control (Temporary Powers) Act. Later, in 1981, cashew became subject to the Essential Commodities Act. Under the Kerala Raw Cashewnut (Procurement and Distribution) Act of 1977, amended in 1981, 1983 and 1988, the trade in raw cashewnut became a state monopoly. Procurement of raw cashewnut was restricted to cooperative societies and appointed private subagents. The raw cashewnut was allotted to the 266 registered factories in Kerala. Each year, furthermore, Kerala's state government was supposed to announce district-wide fixed prices valid for the entire harvest season. Under the Essential Commodities Act, the trade in raw cashewnut over state borders became illegal. Cultivators and non-appointed traders were also forbidden to store more than 50 kg of raw cashewnut. The so-called *state-monopoly procurement* was discontinued between 1984 and 1987, as well as in 1993 and in 1995. In those years, as well as at the beginning of each season until March, Kerala's government neither announced fixed prices nor organized the marketing, and permitted raw cashewnut to be sold in the open market.

Official objectives for introducing the state-monopoly procurement were:

- to ensure remunerative prices for cashew growers;
- to give incentives for cashew cultivation;
- to protect the workers in Kerala's cashew industry.

However, the Raw Cashewnut Act actually was primarily a measure to ensure the supply of reasonably priced raw cashewnut to the 34 public-sector factories in Kerala (Kannan 1981: 85, 131). Particularly in the mid-1970s, these public-sector factories faced difficulties in purchasing domestic raw cashewnut because of stiff competition from private factories in and outside Kerala. The private factories were able to offer higher prices to cashew growers because they did not pay the statutory minimum wage to their workers.

Farm-gate prices for cashew have increased in Kerala since the implementation of Kerala's Raw Cashewnut Act (i.e., from Rs. 2.40 per kg in 1975-76 to Rs. 21.30 per kg in 1993-94). Yet, prices did increase both in years with fixed prices and in years without fixed prices. Moreover, the objective of protecting jobs in Kerala's cashew industry was not achieved under this act. Rather, raw cashewnut was "smuggled" out of Kerala because the production costs (particularly the labor costs) and often also the sales tax are lower in other Indian states. For the "smuggled" nuts, private cashew manufacturers and traders are able to give the cashew growers in Kerala high prices above the fixed price. Therefore, Kerala's government actually only procures a relatively small amount of the raw cashewnut produced in Kerala – in 1994, it was only 27 percent, for example (Cashew Bulletin 1995, no. 2: 2).

Consequently, Kerala's government faced a dilemma. When the farm-gate price is too low, cultivators tend to sell to smugglers so that only small quantities are procured in Kerala. When the farm-gate price is too high, public-sector factories run at a loss. Meanwhile, private cashew manufacturers try to substitute cheaper imported raw cashewnut for the domestic nuts in order to remain competitive in the world market (Deepa 1994: 56).

In April 1995, after a group of cashew growers sent in writ petitions, the Kerala High Court quashed the Raw Cashewnut Act and stopped the application of the Essential Commodities Act to cashew. According to the court, these acts violated principles of India's constitution, which gave individuals the right to practice any profession or to carry on any occupation or trade or business (The Hindu 19.4.1995; Indian Express 19.4.1995). However, Kerala's left government, which came to power in 1996, is considering the reintroduction of state-monopoly procurement and of fixed prices for raw cashewnut.

7.2   Local Trends in Cashew Cultivation

This case study concentrates on the rural region of Mattanur-Iritty in the midland of Kannur District, North Kerala. Mattanur and Iritty are the commercial centers of a region comprising Mattanur Panchayat and Keezhur-Chavassery Panchayat. These two panchayats – in the following simply referred to as Mattanur-Iritty – cover an area of approximately 100 square kilometers, inhabited by almost 75,000 people (roughly, 16,000 households). Compared with Vazhakulam, Mattanur-Iritty is a "backward" region. Unemployment and underemployment are very widespread among the rapidly growing population. Agriculture remains by far the most important source of income. For casual workers, laterite-brick mining has become another important source of income. Especially in Iritty, some people are also employed in the trade of agricultural commodities and in related activities such as transportation and loading.

Mattanur-Iritty lies in the foothills of the Western Ghats. In the northeast of the region, steep and rocky hill tracts and in the south and west, gentle slopes predominate. Relatively few low-lying plains (wetland) intercept the hilly upland area, where lateritic soils are widespread. Soils are generally not very fertile there. On most upland, the natural vegetation is light forest *(kadu)*, bush and grassland. Only near Iritty are upland soils sufficiently moist and fertile so that dense rainforest *(vanam)* could once grow naturally there. In today's land use in Mattanur-Iritty, coconut groves and cashew plantations predominate among other forest-like cultivation forms. Current land-use changes are comparatively moderate. Since 1970, as elsewhere in Kerala, wetland paddies have been converted into dryland, where coconut, arecanut and banana are grown. Furthermore, farmers have replaced a lot of kadu forest with coconut-based and arecanut-based mixed gardens, and rubber and cashew plantations. Consequently, agricultural production has been intensified on upland. However, some parts of Mattanur-Iritty have not been brought under intensive cultivation yet, but are still covered with kadu forest.

Cashew trees have grown "spontaneously" in Mattanur-Iritty for centuries. In the last century, people did not use the cashew tree very intensively. They picked up some cashew apples from the ground, and roasted and consumed only a few cashewnuts. In the 1930s, however, cashew cultivation increased significantly, as local informants remember. On a previously barren, 300-acre-large high-plateau, for example, a Christian migrant started a cashew plantation. At this time, cashew also became a commercial crop that was traded to colonial trading centers on the nearby Malabar coast. In the marketplace in Iritty, furthermore, agents of cashew-processing factories coming from Mangalore (South Karnataka) and Tellicherry started to purchase raw cashewnut. Since the 1940s most cashew agents have come from South Kerala (Kollam). From 1930 to 1980, cashew cultivation and trade in Mattanur-Iritty grew steadily. Particularly on so-called "waste" land and in steep areas, new cashew plantations were established. Cashew trees were also planted more frequently in home gardens and extended mixed gardens.

Since the 1980s, cashew cultivation has not expanded in Mattanur-Iritty. However, according to data from the Krishi Bhavan, Mattanur Panchayat, and the Krishi Bhavan, Keezhur-Chavassery Panchayat, in 1995, still about 5,800 acres (23 percent of the total area) were under cashew cultivation in that region. Nearly all households grow a few cashew trees in their home garden. Furthermore, 300 to 600 households (2 to 4 percent) also grow cashew on plantations. About 20 to 30 percent of the cashew trees are harvested by leaseholders, including poor and middle-class people. Plantations, most of which are between two and ten acres, account for about 80 percent of cashew production; the remaining 20 percent is produced in mixed gardens. Table 5 gives an impression of the various crop shifts related to cashew cultivation between 1980 and 1995.

Cultivators cut down only the old, unproductive cashew trees. According to my own estimate, about 40 percent of the cashew trees had to be replanted between 1980 and 1995. Table 5 shows that, of these, cultivators replanted around 65 percent with cashew (two-thirds with traditional varieties and one-third with high-yielding varieties) and replaced the other 35 percent with rubber plantations, coconut groves, mixed gardens and laterite mines, thus reducing the area under cashew by about 14 percent. On the other hand, farmers have expanded cashew cultivation to former waste land by about 4 percent of the area under cashew cultivation. Thus, the net decline of the area under cashew in Mattanur-Iritty since 1980 has been about 10 percent.

Table 5 shows that among the crop shifts away from cashew, replacing cashew plantations with rubber plantations has been the most significant in Mattanur-Iritty. Since 1980, farmers have replaced 15 to 25 percent of old cashew trees with rubber plantations, and 10 to 15 percent with coconut groves and mixed gardens. Particularly the shift to rubber cultivation has accelerated recently. In some parts of the region, the hard laterite soil is used for brick production. Less than 5 percent of old cashew trees have been cut down since 1980 for the purpose of temporarily producing laterite bricks. Considering these crop shifts, cashew cultivation in that region is best described as a situation of both continuity and moderate change that amounts to a limited intensification of land use.

Table 5   Development of Cashew Cultivation in Mattanur-Iritty

	Percentage of Area that was Under Cashew Cultivation in 1980
Continuation of Cashew Cultivation, 1980-1995	*86*
Continued Cultivation of Cashew	60
Old Cashew Replanted with Traditional-Variety Cashew	17
Old Cashew Replanted with High-Yielding Cashew	9
Shifts Away From Cashew Cultivation, 1980-1995	*14*
Shift to Rubber Plantations	8
Shift to Coconut Groves and Mixed Gardens	4
(Temporary) Shift to Laterite Mines	2
Replacement of old Cashew on Kadu Forest	«1
Shifts to Cashew Cultivation, 1980-1995	*4*
New Cultivation of Cashew on Waste Land	4
Net Decline in Area Under Cashew, 1980-1995	**10**

Source: Estimates based on field observations and on interviews with cultivators, traders and local government officers.

Cashew-cultivation methods applied in Mattanur-Iritty have not changed much, except for the application of chemical insecticides since the late 1970s, the partial adoption of high-yielding varieties, introduced in 1981, and, to a small extent, increased fertilizer application. Generally, cultivators let cashew grow with less care than is given to most other crops. Weeding on cashew plantations is not widespread; harvesting remains the most labor-intensive operation, but lasts only for about four months.

7.3 Socioeconomic and Environmental Impact

Only for about 2 to 4 percent of the households in Mattanur-Iritty, the cultivation of cashew on small and big plantations forms a relatively substantial source of income. Generally, rich and middle-class cultivators run the big cashew plantations and gain the most from raw-cashewnut production. For most of these farmers, cashew cultivation is only subsidiary source of income, supplementing returns from rubber cultivation or income from non-agricultural pursuits. Most other households gain a small, though for poor households often significant, income from growing a few cashew trees – in the home garden, on small plantations or mixed gardens. Since cashew does not involve high costs and much labor, poor cultivators are also able to engage in, and profit from, cashew cultivation – provided they have access to land.

However, the income from cashew cultivation is not stable because cashew yields are very sensitive to weather conditions. Moreover, cashew cultivation tends to be comparatively seasonal and unprofitable; but poor cultivators usually lack the necessary capital, labor or suitable land to engage in the cultivation of more remunerative but also more demanding crops such as rubber. Middle-class and rich cultivators, too, cannot instantly adapt their cropping pattern in accordance with changes in agricultural-commodity prices. This is because cashew is a long-standing tree that is often grown on very poor soils, where no other crop would yield satisfactorily.

While cashew cultivation may provide a subsidiary income for rich, middle-class and poor farmers in Mattanur-Iritty, it does not create much additional on-farm employment. Only a very few agricultural laborers are employed on cashew plantations, mainly during the harvest period from February to May. Also, trading and transportation of raw cashewnut do not involve many people. By contrast, the processing of cashew generates much employment. In Kollam, South Kerala, more than 100,000 factory workers, mostly women of the poorest section of the society, are employed in the registered and unregistered cashew factories. This is about 5 percent of the workers engaged in Kerala's secondary sector. However, jobs in the cashew industry, especially in the home-cottage sector, are very low-paid and also have become more casual and less reliable (Deepa 1994).

From an environmental perspective, cashew cultivation, as it is done in Mattanur-Iritty, has a positive impact. In general, the cultivation of permanent crops tends to be ecologically sound. Cashew cultivation, in particular, is suitable for the poor soils that are widespread in Mattanur-Iritty. The most valuable effect of cashew cultivation on the environment is the control of soil erosion. On the steep slopes in the region and under the heavy rains during monsoon, soil erosion would be much higher if the land were used as pasture, for instance. The extensive lateral root system of cashew, the tree cover and the leaf cover on the ground check soil erosion. Local cashew growers also mention that the roots of the cashew tree break hard laterites and loosen the earth, making it more fertile. Furthermore, the cashew tree does neither deplete water resource nor extract scarce soil nutrients excessively. Thus, only a little chemical fertilizer is needed, if at all.

In order to control pests (tea mosquito, stemborer), however, cashew growers apply chemical insecticides *(endosulfan, carbaryle)*, which are very toxic and can be harmful and even lethal to bees and to other non-target insects, birds, fish and even mammals. Carbaryle can produce adverse effects in humans by skin contact, inhalation or ingestion (Extoxnet 1993). Agricultural workers and farmers, in particular, run the risk of exposure to carbaryle while spraying the cashew trees. Fortunately, the dosage of these toxic substances is fairly modest. The application of *neem*-based bio-pesticides (as opposed to chemicals) is also becoming more popular. Chemical pollution of soils, rivers, groundwater and air through cashew cultivation is therefore insignificant. Many birds and, to a lesser extent, mammals still populate cashew plantations. Only very exceptionally are chemical residues to be found in the cashew kernel (Cashew Bulletin 1995, no. 6: 11).

As the crop shifts away from cashew, and agricultural intensification that involves increased use of agro-chemicals, are fairly moderate in Mattanur-Iritty, the environmental consequences of these changes are generally not severe. On the contrary, the modest shift to terraced mixed gardens even has positive environmental effects. For rubber cultivation, furthermore, farmers follow the recommendations of India's Rubber Board to build contour bunds so that massive soil erosion and washing-out of fertilizers are prevented. Besides, the shift to laterite mines involves only a temporary loss of cultivable land. Former mines are filled up and once more planted with cashew or other crops.

Like the pineapple cultivators in Vazhakulam, the cashew growers in Mattanur-Iritty rate socioeconomic concerns higher than environmental ones. Because it is only a subsidiary, seasonal and unreliable business that does not create much local employment, farmers, agricultural workers and off-farm workers regard cashew cultivation as relatively unimportant. Cultivators also have less confidence in cashew cultivation than in rubber or coconut cultivation because of the uncertainty arising from the irregular implementation of state-monopoly procurement. Furthermore, local farmers and traders do not believe that cashew cultivation will become more profitable in future. They think that government policies are biased toward the cashew-processing industry, including the cashew-factory workers. Because cashew growers assume a trade-off between better wages in the cashew-processing industry and remunerative raw-cashewnut prices, they show no solidarity with the deprived cashew-factory workers in Kollam but rather see them as direct rivals. Consequently, unlike India's Directorate of Cashewnut Development or the Cashew Export Promotion Council, most cashew growers are worried less about the low productivity of cashew cultivation than about government policies.

Although they put socioeconomic aspects in first place, cashew growers in Mattanur-Iritty are aware of the positive environmental effects of cashew cultivation, such as the capacity of cashew trees to control soil erosion. Yet, environmentally sustainable cashew cultivation is not practiced out of a sensitivity to ecological processes. Building bunds on rubber plantations or building terraces in mixed gardens are also not driven primarily by ecological considerations, but rather by economic ones. For example, bunds on rubber plantations prevent the washing-out of chemical fertilizers and thus lead to higher productivity. Furthermore, some rubber cultivators simply follow the Rubber Board's recommendations that include the building of bunds. Generally, environmental protection is not a value in itself, but is only considered if it is linked to economic gain.

7.4   Secondary Influence of the Cashew Market

Since 1980, about 75 percent of the cashew cultivation in Mattanur-Iritty has remained unchanged. The few changes that have taken place include new cultivation of cashew on waste land and replanting old cashew plantations with rubber plantations, coconut groves or other crops. Cultivators have made a variety of partial decisions (e.g., whether to replant with cashew or with another crop; whether to replant with traditional varieties or with high-yielding cashew; whether to shift to rubber or to any other crop). Naturally, their decisions differ.

This analysis starts with the cultivators' own explanations for continuing cashew cultivation, for replanting with traditional or high-yielding varieties, or for shifting to other crops. After the brief discussion of individual motives and, particularly, the role of customary practices for continuity and change regarding cashew cultivation, the analysis continues with the consideration of bio-physical factors, with reflections on agrarian relations and crop markets and, finally, with an assessment of state interventions regarding cashew-cultivation technology. As in the pineapple case study, I attempt to establish the relative importance of the crop market in determining agricultural practice.

7.4.1   *The Cultivators' Viewpoint: Reasons for Cashew Cultivation*

Generally, cultivators in Mattanur-Iritty do not attribute great significance to cashew cultivation as opposed to the cultivation of crops such as rubber, coconut, pepper; and, for poor peasants, as opposed to casual wage labor. Consequently, farmers are less inclined than policy-makers or cashew processors to discuss cashew cultivation among themselves. Because growing cashew is comparatively painless, the subsidiary income from this crop is seen as a gift rather than as a return for hard labor. Local informants were surprised that a foreign researcher showed interest in cashew cultivation and wanted to learn their reasons for growing cashew. For their part, they pursue cashew cultivation without much questioning, and often as a customary practice. However, this does not imply that cashew cultivation lacks an economic (or other) denotation. Indeed, when they are asked explicitly, cultivators are able to give reasons for, or to "rationalize," this customary social practice.

The relative continuity of cashew cultivation can be accounted for by the farmers' aversion to cutting down still productive cashew trees. Kerala's farmers would never remove productive cashew trees (or other tree crops with a long gestation period) in order to replace them with other crops that are more remunerative. Since 1980, 60 percent of the cashew trees in Mattanur-Iritty have remained productive. Cashew cultivation has therefore continued without cashew growers having considered alternatives or having taken actual "decisions."

Another sign of the relative continuity of cashew cultivation in Mattanur-Iritty is that, out of the 40 percent of the cashew plantations that have become unproductive since 1980 and thus had to be replanted, farmers replanted 65 percent with cashew again. The main reason of both rich and poor farmers for replanting with cashew trees has to do with soil characteristics. While some cultivators maintained that no other crop but cashew will grow on their land, other cultivators reasoned that their land is most suitable for cashew cultivation, implying that cashew cultivation is most profitable there.

Both rich and poor farmers maintain that cashew cultivation does not require high inputs of labor and capital and still provides a (subsidiary) income. Poor and lower-middle-class peasants, in particular, appreciate that growing cashew does not impose high operating or replanting costs compared with other crops. Moreover, leasing cashew plantations cuts cultivation costs down to zero, and the rent can be used in emergencies. Upper-middle-class and rich farmers, too, value the comparatively low labor and capital requirements of cashew cultivation. They also appreciate the fact that it enables them to avoid annoyances with wage laborers, labor unions, money lenders, etc. Leasing out their cashew plantations reduces annoyance, time, and risks even further.

Another "rationalization" for continuing cashew cultivation is to maintain a diversified agricultural income. Furthermore, particularly poor and lower-middle class farmers find cashew cultivation profitable compared with the other crops they can afford to grow. Raw cashewnut fetches relatively good prices and, unlike fruit or vegetables, for example, is always saleable in Mattanur-Iritty. Rising prices for raw cashewnut have even motivated some farmers to start new cashew plantations on former waste land.

Only one-third of the cashew trees replanted since 1980 have been high-yielding varieties as opposed to traditional varieties. At present, the re-plantation rate of high-yielding varieties is about 40 percent. Many cashew growers did not replant with high-yielding varieties because such grafts were not available or were too expensive for them. As well, both rich and poor cultivators are sometimes unwilling to engage in the little additional work required by the young high-yielding-variety trees. On very poor soils and on rocky land, the cultivation of traditional cashew trees tends to be more suitable than the cultivation of high-yielding trees, which demand more soil nutrients. Some cashew growers also fear that high-yielding cashew needs to be replanted more frequently. Those cultivators who grow high-yielding cashew do so because these trees usually yield more and earlier in the season, when raw-cashewnut prices tend to be high.

The shift away from cashew cultivation in Mattanur-Iritty since 1980 has both general and specific explanations. The most important reason generally given for abandoning cashew cultivation is that cashew is less profitable than other crops. Other reasons include the pronounced season-ality of cashew (i.e., cashew yields only during three to four months while rubber trees yield during eight months and coconut yields year-round); the comparatively high dependence of cashew yields on weather conditions; the uncertainty regarding cashew prices; the complications and uncertainty regarding the marketing system (state-monopoly procurement); the land

ceilings under Kerala's Land Reform Act, which exempts rubber plantations but not cashew plantations from land ceilings; and the relatively low
efficacy of governmental cashew-development schemes.

Particular reasons to shift to an alternative crop differ between poor and
the rich cultivators. Middle-class and rich farmers choose to grow rubber
because rubber cultivation is very profitable and prices for rubber sheets are
continuously increasing. Poor and lower-middle class farmers, on the other
hand, prefer to shift to coconut cultivation and mixed gardens because these
require less initial capital than does rubber cultivation.

Table 6 summarizes the cultivators' "rationalizations" and explanations
for continuing or discontinuing the customary social practice of growing
cashew. The most important arguments for and against cashew cultivation
are mentioned at the top of the table, the least significant ones at the
bottom. Because farmers have differing perceptions of, and opinions about,
several aspects of cashew cultivation, the same point may appear as a
positive to some and as a negative to others. For example, some cultivators
regard raw-cashewnut prices as fairly good while other cultivators regard
the very same prices as low. When interpreting Table 6, it is also important
to bear in mind that farmers usually pursue cashew cultivation as a social
practice that does not require them to make many deliberate decisions.

Table 6   Rationalizations Regarding Cashew Cultivation

Reasons for Cashew	Reasons against Cashew
- also on poor soils, good growth and yield	- pronounced seasonality
- very low capital and labor inputs required to get satisfactory yield	- yields sensitive to weather conditions
- adds to diversification of agricultural cash income	- complications of, and uncertainty with, the marketing system
- relatively good prices	- comparatively low prices
- relatively good marketability	- ineffective governmental cashew-development schemes
- relatively low replanting costs	- land ceilings for cashew plantations

Different aspects of the cashew market (i.e., prices, marketing systems)
appear as reasons for both continuing or increasing cashew cultivation, and
for shifting away from cashew. Overall, Table 6 illustrates the great weight
cultivators attach to bio-physical factors, and the relative unimportance of
marketing considerations.

7.4.2 Interpretation of Individual Motives and the Role of Traditional Practices

That cultivators stick to customary practices helps to explain the relative continuity of cashew cultivation in Mattanur-Iritty. Particularly the reluctance to replace cashew trees that are still productive can be interpreted as a "traditional" practice because farmers never question it. A few cashew growers do not even consider alternatives to cashew cultivation when they have to replant old cashew trees. Therefore, the farmers' rationality and decision-making need to be put into perspective with the importance of traditional practices. However, cashew cultivation has an economic "function" even for growers who are simply following traditional practices, but derive subsidiary cash income from cashew. The importance of traditional practices may slow down changes in cashew cultivation. However, farmers would probably not continue to grow cashew under any circumstances – for example, if cashew cultivation were so unfavorable economically that its pursuit would threaten the cultivators' livelihood and survival. In the long run, moreover, changing circumstances, such as a sustainable alteration in relative prices, may alter traditional practices.

When old cashew trees need to be replaced, cultivators generally take more intentional decisions. Individuals have very diverse motives both for changing and for continuing cashew cultivation. Furthermore, more than a single motive may underlie a particular decision of a farmer regarding cashew cultivation. For replanting with cashew trees again, cultivators have such different motives as minimizing costs, drudgery and annoyance or maximizing profits (if the land is most suitable for cashew cultivation). For intensifying land use (i.e., shifting away from cashew cultivation; adopting high-yielding cashew; starting cashew plantations on waste land), the profit-maximization motive seems more important, particularly among those middle-class and rich cultivators who shift to rubber cultivation. Lower-middle-class and poor cultivators, who tend to shift to coconut groves and mixed gardens rather than to rubber plantations, are motivated equally by a desire to avoid costs and risks, and a desire for greater profits. For growing rubber, status motives may also play a role. Rubber cultivation is prestigious because of its high capital intensity and because officers of India's Rubber Board pay much attention to rubber growers and visit them frequently from the city of Kottayam, which is about 400 kilometers from Mattanur-Iritty.

However, individual motives may not always determine actual practices. Poor peasants, for example, lack resources such as suitable land, sufficient labor and capital that are initially required to change the existing land use or that are needed to cultivate crops other than cashew. Generally, middle-class and rich farmers have more choice: They can *afford* to maximize profits or to minimize annoyance. In fact, the aversion to annoyance is a relatively important subjective motive for middle-class and rich farmers to continue cashew cultivation. However, the particular motives regarding cashew cultivation are not necessarily the same as the cultivators' general motives. For example, a rich farmer who continues cashew cultivation in order to avoid additional annoyances with laborers, labor unions, etc. may have profit-maximizing motives for the cultivation of other crops or in regard to a non-agricultural business.

The partial intensification of land use in Mattanur-Iritty may be linked to value changes toward profit. With the information gathered in the field, however, one cannot determine whether profit-oriented motives have become more important since 1980. Yet, it does seem that the partial intensification of land use in Mattanur-Iritty is influenced less by value changes than by changes in land relations and relative prices for agricultural commodities (see Sections 7.4.5 & 7.4.6). To account for the relative continuity of cashew cultivation, furthermore, one must also look at biophysical conditions and plant characteristics as well as the low-cost nature of cashew cultivation (see Sections 7.4.3 & 7.4.4).

7.4.3 *Continuing Importance of Bio-Physical Factors*

The distinct ability of cashew to grow well on poor soils and to yield for as long as 50 years has motivated farmers to continue cashew cultivation. On the widespread hard laterites and rocky soils in Mattanur-Iritty (and elsewhere in Kannur District and Kasaragod District), cashew trees still grow and yield quite well, while crops such as coconut, rubber or arecanut have ceased to grow or yield. Furthermore, the climate of North Kerala, with its extended and pronounced dry season, is favorable for cashew cultivation but not suitable for many other crops. Rubber cultivation, for example, is less productive in Kannur District than in Kottayam District (288 kg per acre as opposed to 348 kg per acre, according to data supplied by India's Rubber Board for 1992-93), has a longer gestation period in Mattanur-Iritty than in Kerala's main rubber-growing areas (ten years as opposed to seven years) and a shorter seasonal harvesting period (eight months as opposed to year-round).

The soil and climatic conditions in Mattanur-Iritty (as in other hilly parts of Kannur District) lead to relatively high cashew yields and very good quality of raw cashewnut. Consequently, raw cashewnut from Kannur District fetches the highest prices on the world market and is always saleable. During state-monopoly procurement in 1994, for example, the government price for raw cashewnut from Kannur District was between 2 and 6 percent, or between 0.50 and 1.50 Rs. per kg, higher than in other districts of Kerala.

The favorable soil and climatic conditions (including the previous availability of waste land) have been important factors for the expansion of cashew cultivation in Mattanur-Iritty since 1930. However, some bio-physical factors – particularly soils – are not fixed but can be improved (or degraded) through particular land-management practices. Soils can be loosened and, thereby made cultivable for coconut or banana, for example, through mining laterite bricks or through the cultivation of cashew. Moreover, some farmers transform their land with the specific purpose of shifting away from cashew, as the following example illustrates:

> On a steep and rocky area in Mattanur-Iritty, a middle-class farmer owns about four acres land. Before 1991, the soil quality was uniformly low on this land, and the farmer grew cashew trees everywhere. Then, he decided to start a rubber plantation on one acre, expecting that rubber cultivation would be more profitable and less seasonal than cashew cultivation. Other incentives to grow rubber were the Rubber Board's technology extension and subsidies. Before the rubber seedlings could be planted between big rocks, however, the farmer had to remove stones from the land, loosen the earth and build bunds. This was very hard work that took much time.
>
> Today, the young rubber trees still need much care, and inputs such as chemical fertilizers and pesticide. Because of the high costs of transforming the land, as well as of rubber seedlings and fertilizers, this farmer cannot afford to plant more rubber trees. Therefore, he has again replanted another part of his plantation with cashew trees of high-yielding variety. He estimated that replanting with cashew cost him about seven times less than replanting with rubber, not including the costs of his own labor.

This example shows that soils do not completely determine land use. Even on very poor soils, cultivators – provided they want to intensify the land use and provided they are able to invest in land-improvement works – may shift away from cashew cultivation. By contrast, there are also cultivators that continue to grow cashew on fertile soils suitable for the cultivation of more remunerative crops. However, cultivators holding land of different soil qualities tend to grow cashew on the least fertile soils and other crops on the most fertile soils. Generally, bio-physical factors, or the "natural" productivity of cashew compared with other crops, can be used to explain crop patterns only in conjunction with the motives of individual farmers and changes in relative prices and costs.

7.4.4 Cultivation Costs

The cultivators' own accounts indicate that low planting or replanting costs, operating costs and labor requirements are the main reasons for growing cashew. Table 7 summarizes the annual operating costs for relatively intensive cashew cultivation on land owned by the cultivator. However, these costs can only be roughly estimated and vary widely among cultivators.

Table 7 shows that total annual costs for intensive cashew cultivation are about Rs. 4,200 to Rs. 6,000 per acre, while average returns may vary between Rs. 5,000 and Rs. 15,000 per acre. In the planting year (every 25 to 50 years), cultivation costs are about Rs. 1,300 per acre higher. Another Rs. 1,500 to Rs. 3,000 must be added if the trees are taken on lease for the harvest. On the other hand, planting and operating costs can be more than 50 percent lower if the cashew grower obtains government subsidies (see Section 7.4.7). Moreover, when the cashew grower does not apply fertilizers and uses only family labor for the harvest, monetary operating costs are close to zero. Compared with competing crops such as rubber, coconut or arecanut, the cultivation costs for cashew are very low.

Table 7   Annual Operating Costs for Intensive Cashew Cultivation

Type of Costs	Rs. per Acre (range)		
Labor (mainly for harvesting)	3,000	-	4,000
Chemical and Organic Fertilizers	600	-	1,200
Chemical Pesticides	600	-	800
Total Costs	4,200	-	6,000
Average Returns	*5,000*	-	*15,000*

Source: Own estimates based on information from cashew cultivators in Mattanur-Iritty.

The low requirements of cashew cultivation not only imply low cultivation costs in monetary terms, but also little annoyance with agricultural workers, labor unions, banks, moneylenders, etc. Farmers prefer labor-saving crops such as cashew not only because agricultural wages have increased rapidly but also because the interaction with wage laborers often means trouble. Most regular work on cashew plantations can be done with family labor, which shows gender differentiation: while men apply fertilizers and insecticides to the cashew trees, women and children collect the raw nuts from the ground. During the harvest season, raw cashewnuts are collected daily, so that they do not rot. Additional labor is hired for harvesting on big cashew plantations of more than five acres. Furthermore, hired labor is employed for replanting and other occasional work on cashew plantations.

The few laborers who work on cashew plantations tend to be submissive. The hired cashew-harvest workers are usually women and children, who, unlike male workers such as coconut or arecanut climbers, are generally not organized in labor unions, which are exceptionally strong in Kannur District. Therefore, the cashew-harvest workers' bargaining power is comparatively low, and labor disputes related to cashew harvesting are rare. Female workers are paid only about Rs. 20 to Rs. 30 for a half-day's work. Children, who may be employed during school holidays, earn less. Furthermore, the low-cost and low-input nature of cashew cultivation leaves cashew growers not only quite independent from hired labor and labor unions, but also from external financial resources and external supplies of agricultural inputs.

The general independence from other agents keeps down the transaction costs for the production of raw cashewnut. Therefore, even though profits may be lower than for other crops, cashew cultivation is still attractive for middle-class or rich cultivators, who appreciate the little annoyance involved in growing cashew. For the poor, on the other hand, the low monetary requirements are an important reason to grow cashew.

7.4.5 Land Reforms, Land Ownership and Leasing

Mattanur-Iritty, where the majority of people rely on agricultural income, has a high population density of nearly 750 people per square kilometer. Among a growing population, land partitions by inheritance render operational landholdings still smaller. Furthermore, landownership remains relatively uneven, although land transfers under Kerala's Land Reform Act of 1970 have been substantial in that region, where the land was previously in the hands of a very few landlords. Hundreds of former tenants have become landowners and thousands of agricultural-worker families have become owners of their house-compound land. On upland, these new small-holders have often intensified the existing land use. For instance, they gradually replaced kadu forest with terraced, mixed gardens. To a lesser extent, they also replaced cashew plantations with coconut and other crops.

Although land was expropriated from big landlords under the land re-form and redistributed to landless, very poor families, the former landlords gave away only those lands that were relatively infertile or remote and that had not been used very intensively (e.g., waste land, kadu forest, cashew plantations). In order to avoid expropriation, some landlords replanted cashew plantations with rubber plantations, which were exempted from the land-ceiling provisions of Kerala's Land Reform Act. Moreover, landlords also bypassed the land ceilings by formally partitioning land among family members: although the land reform of 1970 imposed land ceilings of 15 to 20 acres, bigger cashew plantations, which may officially belong to different people but are managed by a single person, extend to 50 acres. Furthermore, the very poor new smallholders, unlike the slightly better-off former tenants, were not able to intensify the land use because of indebted-ness and lack of farming experience. Poverty and dependence led to un-sustainable practices, as the following example illustrates:

> In the 1930s, a migrant from central Kerala bought 300 acres of rocky waste land of a high plateau, where he planted cashew trees. On this land, his family produced raw cashewnut for nearly half a century. Under the land-redistribution provision of Kerala's Land Reform Act, this land was distributed to more than 100 landless families in the mid-1970s. Yet, most of the new landowners did not continue cashew cultivation but cut down the cashew trees, the wood of which was sold in order to repay old debts and to meet urgent expenses. By the mid-1980s, most of these new landowners had vacated the land again. Today, only a few cashew trees are still grown here. On most parts of this high plateau, the hard laterite soil is mined to produce bricks.

Changes of land ownership under Kerala's land reform have had an impact on agriculture. Many new landowners have been more interested than the former landlords in intensive agriculture. The better-off new owners have been able to intensify the land use; the very poor peasants have been forced to apply unsustainable land-use practices for short-term gain. Eventually, both processes may have contributed slightly to the decline of cashew culti-vation in Mattanur-Iritty.

Land prices in Mattanur-Iritty have increased rapidly for the past decade. In 1995, agricultural land in Mattanur-Iritty cost about Rs. 300,000 to Rs. 600,000 per acre. Since land is often regarded primarily as an asset, even people who are not interested in cashew cultivation are reluctant to sell their land. Instead, they often lease the cashew plantation for the harvest. Doing so enables the plantation owner to pass the main cultivation costs (i.e., the harvesting costs) onto the leaseholder. Unlike in the case of pineapple cultivation in Vazhakulam, the leaseholders are not only middle-class people (including small cashew traders) but also relatively poor people. Usually, leasing contracts are only for one cashew harvest. The rent varies between Rs. 1,500 and Rs. 4,000 per acre and has to be paid in advance, at the beginning of the harvest season in February. (In many other places in Kerala, cashew plantations are also grown on state-owned "forest land" or "reserve forest." These plantations are leased for harvesting to contractors who employ casual wage laborers, including children.)

Furthermore, unlike rubber plantations, which are leased only for "slaughter tapping" (extraction of as much latex as possible in the last productive year of rubber trees), cashew plantations are not easily damaged by inappropriate cultivation methods. Therefore, cashew-plantation owners do not need to supervise the leaseholders' practices. Because only the right to harvest is leased, leasing cashew plantations is also not in conflict with Kerala's Land Reform Act, which prohibits only land tenancy. Thus, there is no danger that former leaseholders are able to claim ownership of cashew plantations.

There are also a few instances of *reverse tenancy*; that is, some poor smallholders lease cashew-harvesting rights to richer farmers. For poor cashew-plantation owners, occasional leasing is important to get immediate money in situations when it is urgently needed (for a wedding, dowry, medical care, etc.). For this reason, they also accept very low rents. This may signify a "simple reproduction squeeze," which, however, does not result in resource overuse or in unsustainable practices.

Generally, the possibility and ease of leasing cashew plantations dissuades particularly annoyance-averse, middle-class and rich farmers from shifting away from cashew cultivation to potentially more remunerative crops such as rubber. This may have contributed slightly to the relative continuity of cashew cultivation in Mattanur-Iritty.

7.4.6 Prices and Controls in the Cashew Market

A market for raw cashewnut has existed for more than half a century in Mattanur-Iritty, and marketing opportunities for cashew growers have not changed much since 1980. Such changes that have occurred in the established cashew market relate to changing prices and to the alternating implementation of state-monopoly procurement and open marketing of raw cashewnut.

Farm-gate prices for raw cashewnut in Mattanur-Iritty have generally increased since 1980 (see Fig. 10). Since then, also, the global demand for cashew kernel has grown. However, changes in the raw-cashewnut price in Mattanur-Iritty do not always correspond to changes in global demand or price. For example, between April 1986 and April 1987, the market price for raw cashewnut in Mattanur-Iritty fell from Rs. 16 down to Rs. 13 per kg while the cashew-kernel price in the New York market rose from US$ 2.68 to US$ 3.32 (Cashew Export Promotion Council 1994). Such discrepancies between the local and global market trends are due to the policies of Kerala's government, which fixes the price of raw cashewnut in years of state-monopoly procurement, or to the cost calculations of India's cashew-processing factories in years of open marketing.

When determining the market price, India's cashew-processing factories consider numerous interdependent factors, many of which go beyond national boundaries. Important determinants include the global demand for cashew kernel, relative prices for cashew kernel and other edible nuts in international nut markets (London and New York), existing kernel stocks, cashew-kernel production in competing countries, weather conditions and export policies for raw cashewnut in cashew-growing countries and the subsequent availability of raw cashewnuts for import, India's import and export policies, domestic supply of raw cashewnuts, and relations with factory workers. During years of state-monopoly procurement, the prices for raw cashewnut are subject to political decisions, which not only consider global demand and supply but also party politics and political affiliations in Kerala. Congress-led governments, which are sympathetic toward farmers, tend to announce higher raw-cashewnut prices than Communist-led governments, which are more supportive of the cashew-factory workers.

The cultivators' own perception of, and response to, changing raw-cashewnut prices are ambiguous. Farmers in Mattanur-Iritty mention raw-cashewnut prices as reasons for both growing cashew and for shifting away from cashew cultivation. The generally increasing raw-cashewnut prices may have motivated farmers to grow cashew on former waste land. At the same time, lower *relative prices* of cashew compared with crops such as rubber and coconut may also have led farmers to shift away from cashew.

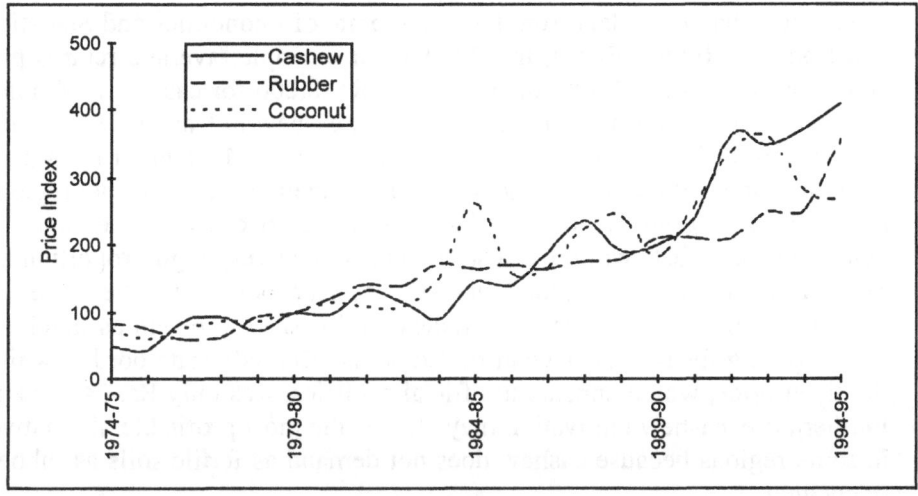

Fig. 10 Prices of Cashew and Competing Crops
Sources: Cashew and coconut prices: Directorate of Economics and Statistics (Statistics for Planning, various issues). Rubber Prices: Rubber Board, cited in Directorate of Economics and Statistics (Statistics for Planning, various issues). Index: 1979-80 = 100 (cashew: Rs. 5.73 per kg; coconut: Rs. 1.16 per nut; rubber (RMA-4): Rs. 10.17 per kg).

Fig. 10 shows no clear long-term trends regarding the relative price of cashew, rubber and coconut. Since 1991, however, relative prices of cashew and rubber have changed very much in favor of rubber. This may have motivated farmers in Mattanur-Iritty to shift from cashew to rubber cultivation. However, farmers also made the same crop shift between 1989 and 1991, when relative prices were changing in favor of cashew. Relative prices of cashew and coconut, moreover, have once more changed in favor of cashew since 1992, but farmers in Mattanur-Iritty still shift from cashew to coconut cultivation rather than the other way round. This indicates that short-term changes in relative prices may have a relatively insignificant influence on cashew cultivation. Bio-physical factors, and the low cost of cashew cultivation, seem more important to explain why farmers continue to grow cashew. As for the shift away from cashew cultivation, this can be explained best by factors such as the pronounced seasonality of cashew and its sensitivity to weather, which makes the annual yield unpredictable.

Since the late 1970s, returns from cashew cultivation in Kerala have generally been lower than from rubber cultivation. The recent hike in rubber prices has further enhanced this general phenomenon. According to calculations based on data from the Directorate of Economics and Statistics and from the Rubber Board, in 1994-95, for example, average returns per acre were about Rs. 12,000 for rubber and Rs. 8,200 for cashew in Kerala. However, because productivity and prices for cashew in Kannur District are higher than the Kerala average, cashew cultivation in Mattanur-Iritty is probably not unprofitable in comparison with competing crops. In that region, raw cashewnut prices tend to be higher not only because of the superior quality of local cashewnut, but also because a relatively high proportion of raw cashewnut can be traded across the near border to the state of Karnataka. In 1994, for instance, traders paid cashew growers as much as Rs. 27 per kg for raw cashewnut traded across Kerala's state border, while the fixed price, which appears in official statistics, was only Rs. 24 per kg. Furthermore, cashew cultivation may also be the most profitable alternative in some regions because cashew does not demand as fertile soils as rubber or coconut.

Considering the local peculiarities, a cashew grower in Mattanur-Iritty may achieve annual returns of Rs. 5,000 to Rs. 15,000 per acre (my own estimate based on information of cashew growers and traders). An additional Rs. 6,000 to Rs. 9,000 per acre can be gained through the sale of the wood of old cashew trees. The returns vary so much because the yields are very sensitive to weather conditions and also depend on soils, cultivation methods, planting material, etc. On a mature cashew plantation, the annual yield varies between 120 kg and 900 kg per acre – a much higher variation than for rubber or coconut. Therefore, the income from cashew cultivation is relatively unstable and unpredictable. This forms another incentive to shift away from cashew to rubber or coconut cultivation. Compared with the annual variation of cashew yields, annual and seasonal fluctuations of cashew prices are relatively minor. Seasonal price variation is further reduced under state-monopoly procurement that fixes prices valid for the whole harvest season. Fixed and support prices reduce the farmers' uncertainty, but the state-monopoly procurement also created problems for cashew growers in Mattanur-Iritty.

Fig. 11 shows the marketing channels for raw cashewnut from Mattanur-Iritty under state-monopoly procurement and under open marketing, with their relative importance.

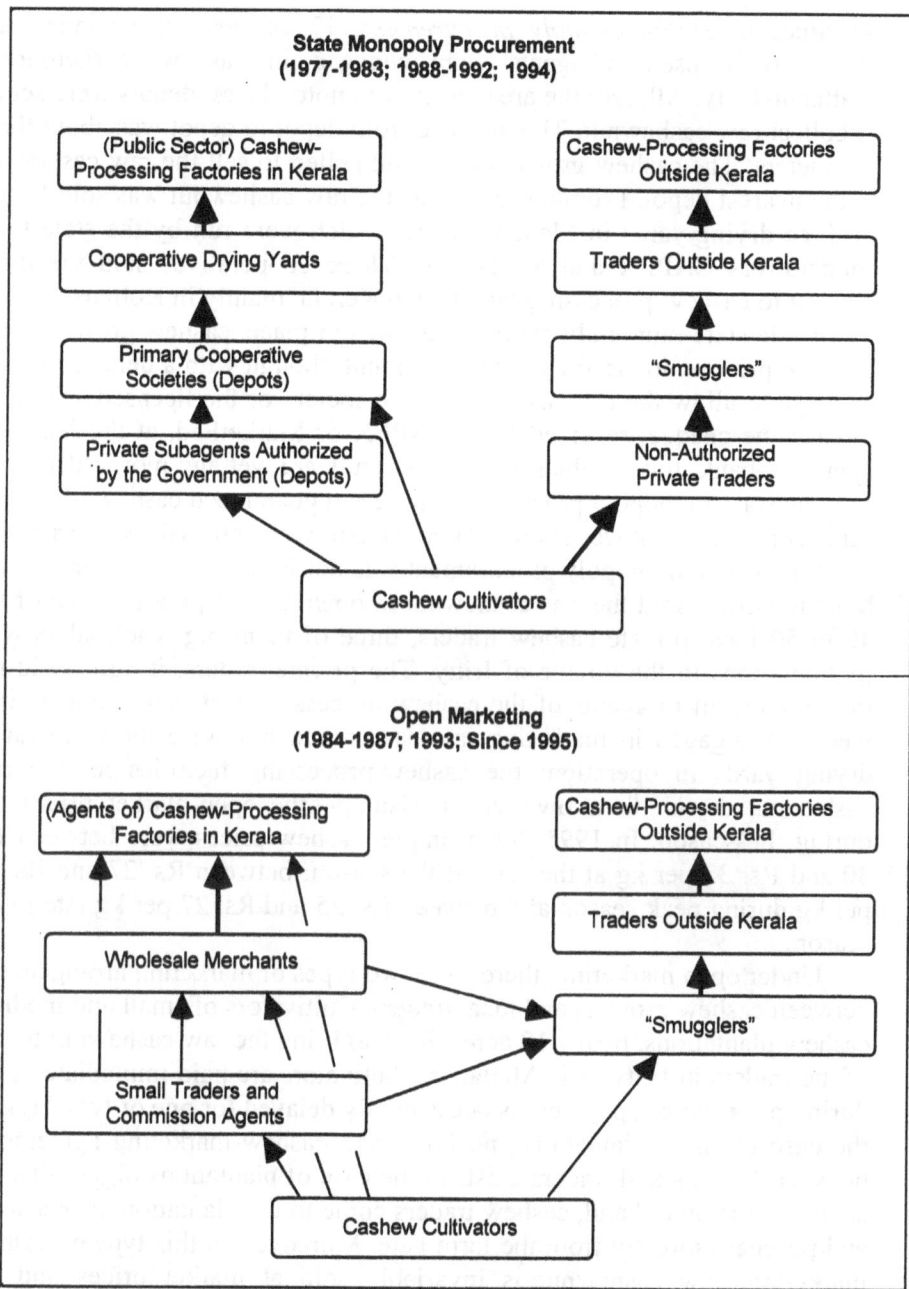

Fig. 11 Marketing of Raw Cashewnut in Mattanur-Iritty
Source: *Based on information from cultivators and traders.*

At times of *state-monopoly procurement*, 17 cooperative societies and about 30 licensed subagents were engaged in cashew marketing in Mattanur-Iritty. All over the area, even in remote places, depots were set up to collect raw cashewnut. The distance from depot to depot was about three kilometers. The cashew growers were compelled to sell the raw cashewnut to the nearest depot. From these depots, the raw cashewnut was sold to one of three drying yards in Mattanur-Iritty, which were run by the state-level cooperatives CAPEX and MarketFed. These cooperatives sold the dried raw nut to cashew-processing factories in Kerala, mainly in Kollam.

While state-monopoly procurement was in place, cashew growers were paid the price announced by the government, though with a delay of two to six days to allow the primary cooperative society or the licensed sub-agent to cash the cheque, received from CAPEX or MarketFed, at the local co-operative bank. Late in the season, when nuts are wet and more difficult to sell, the state-monopoly procurement program guaranteed cashew growers a support price, which was about 10 percent below the normal fixed price.

When state-monopoly procurement was not in force, cashew growers in Mattanur-Iritty sold the raw cashewnut at open-market prices to one of the 40 to 50 local private cashew traders, three of them big wholesalers with trading shops in the village of Iritty. The private traders, in turn, sold the raw cashewnut to agents of the cashew-processing factories. Cooperatives were not engaged in marketing raw cashewnut, nor were the cooperative drying yards in operation: the cashew-processing factories let the raw cashewnut dry in their own yards. Usually, the open-market prices fell during the season. In 1995, for example, cashew prices were between Rs. 30 and Rs. 32 per kg at the start of the season; between Rs. 27 and Rs. 29 per kg during peak season and between Rs. 25 and Rs. 27 per kg late in the season.

Under open marketing, there were two types of marketing arrangements between cashew growers and local traders. Cultivators of small and medium cashew plantations, below 10 acres, have to bring the raw cashewnut to one of the traders in Iritty or in Mattanur. Cultivators are paid immediately, but during peak season, payment is occasionally delayed for one or two days. In the case of small plantations, no long-term cashew-marketing agreements between farmers and traders exist. In the case of plantations bigger than 10 acres, on the other hand, cashew traders come to the plantation once a week and purchase directly from the farm gate. Moreover, in this type of cashew marketing, raw cashewnut is invariably sold at market prices and the payment is not delayed for more than two days. A few of the bigger cashew growers sometimes manage to obtain weekly cash advances from wholesalers or factory agents. This case study cannot confirm observations of Kannan (1981) that before the start of the harvest season in March, cashew growers in North Kerala received loans and cash advances from wholesalers and cashew-processing factories in order to finance the relatively costly collection of raw cashewnut.

Both in years of state-monopoly procurement and in those of open cashew marketing, raw cashewnut from Mattanur-Iritty has also been traded across the nearby state border to Karnataka. This illegal trade accounts for about 50 to 75 percent of raw cashewnut from Mattanur-Iritty, although police patrols frequently check jeeps and farms for smuggled raw cashewnut. Furthermore, on the state highway to Mysore in Karnataka, there are border check-points staffed by Kerala's revenue department.

The irregular implementation of state-monopoly procurement adds substantially to uncertainty among cashew growers and traders. In some years, Kerala's government did not introduce state-monopoly procurement at all. In the years when it did introduce state-monopoly procurement, it was erratic in its timing of the announcement. Moreover, prices were fixed not at the beginning of the cashew-harvest season, but without advance notice at any time between mid-February and mid-April.

To a certain extent the state-monopoly procurement is also responsible for seasonal price trends because private factories in and outside Kerala are anxious to ensure enough raw cashewnut before the implementation of state-monopoly procurement, which gives priority to public-sector and cooperative factories. Fixed prices may also aggravate the effects of a bad harvest by preventing the rise in market price that would help to counterbalance the temporary reduction in the production of cashewnut. One must conclude that fixed prices, although they are designed to lower the cultivators' transaction costs and increase their security, do not always have beneficial consequences for those they are meant to help.

Except for the support price for wet raw cashewnut and the reduction of transportation costs made possible by the narrow grid of collection depots that reaches into remote areas, the state-monopoly procurement offers little to cashew growers. With or without state-monopoly procurement, raw cashewnut can be sold at any time; and the cashew growers can store and dry the nuts themselves. Moreover, the private cashew market is not interlocked with factor markets and not more "exploitative" than state-controlled marketing.

While it has only a few positive qualities from the farmers' viewpoint, the state-monopoly procurement has serious negative effects, particularly for big cashew growers. In years of state-monopoly procurement, police patrols prevented cultivators from selling to the trader who offers them the best price. Furthermore, the police's revenue squad frequently checked big farms, and jeeps, for illegally stored or transported cashew weighing more than 50 kg – less than a big farmer's one-day harvest. In order to avoid seizure of raw cashewnut and a subsequent lawsuit, farmers bribe the police, whose primary objective is, according to local opinion, to benefit from, rather than to obstruct, the trade of raw cashewnut across the border to Karnataka. There is even an informal agreement among the officers of the revenue squad to rotate the postings to Mattanur-Iritty and to other important cashew-growing areas near the state border so that each officer is able to benefit from the "extra payments" at some time. Despite the amount that traders have to pay as bribes, cashew growers can still be offered Rs. 1 to Rs. 3 per kg more for "smuggled" cashewnut than for cashewnut sold to factories in Kerala. Furthermore, the complicated payment system, which results in delayed payment for cashew growers, makes the transaction costs under state-monopoly procurement higher than under a system of open marketing. Finally, big cashew growers accuse the cooperatives involved in cashew marketing of inefficiency and of dishonest practices, as the following case illustrates:

> Because the cooperative banks were short of cash money, a big raw-cashewnut producer in Mattanur-Iritty had to wait for six months in 1992 before getting paid for the produce of the whole season. Furthermore, the cooperatives usually rejected 10 percent of the procured raw cashewnut of the same farmer with the pretext that those nuts were damaged. However, the proportion of damaged nuts normally is only about 3 percent. According to this farmer, the cooperative workers put a big part of the rejected raw cashewnut into their own pockets.

For all these reasons, big cashew-plantation owners, in particular, oppose state-monopoly procurement. Cashew growers in Mattanur-Iritty also think that fixed prices for raw cashewnut bear no relation to the international market-price for cashew kernel and they feel exploited both by cashew manufacturers and by unionized cashew-factory workers in Kollam. Consequently, cashew growers organized in Kannur's Cashew Growers' Association have held demonstrations against the state-monopoly procurement and have sent writ petitions to the Kerala High Court (see Section 7.1). Apart from the cancellation of the Raw Cashewnut (Procurement and Distribution) Act, the cashew growers' demands included free trade for raw cashewnut within Kerala and across state borders, the introduction of a floor price of Rs. 35 per kg of raw cashewnut, and the granting of licences for additional cashew-processing factories in Kannur District.

In sum, Kerala's state-monopoly procurement of raw cashewnut is characterized by erratic implementation, inefficiencies and corruption, all of which increase the transaction costs of cashew marketing compared with open marketing. While the transaction costs for the *production* of raw cashewnut are comparatively low, the transaction costs for cashew *marketing* are high compared with the private and open marketing of rubber and with coconut marketing, which involves floor prices but also competition between cooperatives and private traders. By implementing fixed raw-cashewnut prices, moreover, Kerala's state government failed to reduce the risks to farmers engaged in cashew cultivation substantially. All in all, uncertainties related to state-monopoly procurement and its irregular implementation may have had a small negative impact on the development of cashew cultivation in Mattanur-Iritty.

7.4.7 State Interventions Regarding Technology

Unlike in the case of pineapple cultivation, there are comprehensive government-sponsored cashew-development schemes. Under the present scheme of the Eighth Five-Year Plan (1992-97), the Government of India allocated Rs. 480 million (about US\$ 16 million) for the Centrally Sponsored Cashew Development Programme. Generally, the production of raw cashewnut is expected to increase from about 350,000 tonnes in 1992 to an ambitiously targeted 600,000 tonnes in the year 2000 (Directorate of Cashewnut Development 1992).

Kerala's Department of Agriculture and its local Krishi Bhavans cooperate with India's Directorate of Cashewnut Development to implement the Centrally Sponsored Cashew Development Programme. In Mattanur-Iritty, there are currently three major cashew schemes. For cultivators that own less than five acres of land, cashew grafts (high-yielding varieties) are distributed free under the Cashew Area Expansion Scheme and the Cashew Replantation Scheme, the latter of which also involves 50 percent subsidies for fertilizers, plant-protection chemicals and general replantation costs. Under the Intensive Production Technology Scheme, furthermore, small, as well as big, cashew growers obtain subsidies during three years in order to improve yields on existing plantations.

Despite these substantial government subsidies, fairly few cultivators in Mattanur-Iritty have adopted intensive cultivation methods (except for the use of insecticides). Cashew growers are reluctant to apply intensive cultivation methods because these involve additional costs and labor. Consequently, the application of intensive cultivation methods would neutralize the major advantage of cashew cultivation: its low cost.

The slow adoption of high-yielding cashew can be partly attributed to the disincentive created by the extra requirements of the young trees: Young, high-yielding cashew trees require more fertile soils, more fertilizer, better plant protection and shading but they start to yield earlier. Six-year old trees already give full yields, whereas traditional-variety trees produce full yields only after 15 years. The high-yielding varieties raise productivity significantly, but productivity drops already after 25 to 30 years, whereas the traditional varieties may still give full yields after 50 years. Generally, cashew growers have not yet been able to gain much experience with high-yielding cashew and the recommended cultivation methods, and are therefore skeptical regarding the promised yields. Furthermore, neither the state (in the shape of India's Directorate of Cashewnut Development, Kerala's Department of Agriculture and the local Krishi Bhavans) nor private nurseries have been capable of producing, selling and distributing vegetatively propagated high-yielding-variety cashew seedlings in sufficient quantities and at the right time.

The new technologies neither suit the bio-physical conditions in Mattanur-Iritty nor the particular situation of the poor and middle-class, part-time cashew growers. The high-yielding varieties of cashew developed so far are not suited to the very poor soils that are widely used for cashew plantations in Mattanur-Iritty. Also, the new varieties of cashew and the cultivation methods recommended by the experts have failed to mitigate the extreme sensitivity of cashew to the weather. Moreover, high-yielding cashew has a comparatively short, pronounced harvest period of two to four weeks. The genetically diverse, cross-pollinated traditional-variety trees, on the other hand, do not all flower at the same time and harvesting may be over four to five months. This implies that high-yielding cashew reduces the harvesting period, making cashew cultivation even more seasonal (see Fig. 12). But poor and middle-class, part-time cashew growers favor equal, relatively small yields over a long harvest period, so that harvesting can be done entirely with family labor. This problem could be mitigated by planting various varieties that have different flowering and harvesting times. However, the Krishi Bhavans in Mattanur-Iritty provide only two varieties, both of which are early yielding. The advantage of these varieties is that they bear ripe nuts between late-January and March, when prices tend to be higher and rainfall less likely to do damage than later in the harvest season, when the nuts get soaked with water.

Generally, cashew-development schemes involve fewer incentives (lower subsidies, less technical support, less credit) than other crop-specific development schemes, notably for rubber cultivation and, more recently, also for coconut cultivation. Moreover, the Rubber Board, especially, has been more efficient and effective than the Directorate of Cashewnut Development in promoting expansion and intensification of agriculture. Therefore, the technological improvements and the activities of state agencies have failed to provide effective incentives that would motivate farmers to grow more cashew and thus to overcome the stagnant condition of cashew cultivation in Kerala.

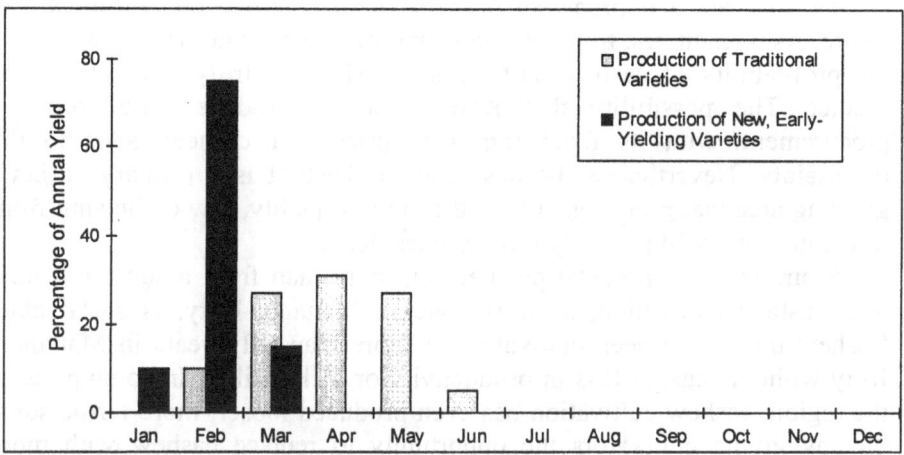

Fig. 12   Monthly Cashew Production
Source: Estimates based on information from cashew cultivators in Mattanur-Iritty and from
the State Farm, Aralam.

7.5   Sustainability, Markets and Cashew Cultivation

The cultivation of cashew does not have much impact on local socio-economic development: it does not create much on-farm employment and is only a subsidiary source of income for cultivators in Mattanur-Iritty. Yet it offers even poor farmers a chance to diversify cash income. Also, the market access is as good and open for poor as for rich cashew growers. But the cashew market failed to encourage farmers to intensify cashew cultivation.

For the cashew industry and for industrial employment, on the other hand, the domestic production of cashew is becoming ever more important as the availability of imported raw cashewnut decreases. However, in Mattanur-Iritty as elsewhere in Kerala, the growth of raw-cashewnut production is limited. An expansion of the area under cashew cultivation is difficult to achieve because only little waste land is still available and because, on better soils, cultivators tend to grow more remunerative crops than cashew. Moreover, cashew growers tend to be reluctant to invest in more intensive cashew cultivation.

In future, the world demand for cashew kernel is likely to increase further. To meet the rising demand, cultivation and processing of cashew are also on the increase in countries other than India. This means that Indian cashew growers and processors can expect increased competition in the future. As a result, the trend of kernel and raw-cashewnut prices, as well as the profitability of cashew cultivation in Mattanur-Iritty, are difficult to predict. The possibility that Kerala may reintroduce state-monopoly procurement, with its fixed farm-gate prices for cashew, adds to the uncertainty. Nevertheless, because Kannur District is a primary cashew-growing area that produces nuts of the highest quality, raw cashewnut from Mattanur-Iritty will probably remain saleable.

From an environmental perspective, more than from a socioeconomic one, cashew cultivation, as it is done in Mattanur-Iritty, is sustainable. Cashew trees have been cultivated for more than fifty years in Mattanur-Iritty without causing loss in productivity or soil fertility. In some parts of the region, cashew cultivation has even produced looser, more fertile soils, thereby giving cultivators the opportunity to replace cashew with more remunerative crops that have higher soil requirements. Furthermore, cashew cultivation does not pollute the soils and rivers with agro-chemicals, or damage the environment in other ways. Processing cashew also does not create severe environmental problems because it is predominately manual, does not require many external inputs, and does not produce much waste. The energy requirements for roasting, for example, are met by burning old cashew shells.

The cultivation of cashew on former waste land in Mattanur-Iritty has proceeded in response to rising demand from India's cashew-processing factories since the 1930s. In this way, the crop market contributed indirectly to the spread of environmentally sustainable agriculture. Since the early 1990s, the cashew market has also influenced the environmental soundness of cashew cultivation more directly. Under pressure from consumer organizations, industrialized countries have introduced strict regulations regarding the quality of food products. Consequently, imported cashew kernel is checked for chemical residues. If contaminated cashew kernel is detected, the culpable processor-exporter will find it difficult to remain in the market. Thus, exporters conduct pre-shipment quality controls. Also, India's Cashew Export Promotion Council, which represents the cashew processors-exporters, informs and warns both cashew growers and processors about the use of harmful chemicals. The inappropriate use of DDT and other harmful pesticides for cashew cultivation and storage have all but vanished. The Council even recommends a total ban on the application of DDT and other harmful pesticides in India (Cashew Bulletin 1995, no. 6: 10-11; Indian Cashew Journal 1994, no. 3: 18). Moreover, the demand for organic cashew is increasing. As a result, a few cashew plantations that use no chemical fertilizers or insecticides have been started recently in India and elsewhere (Market Asia Index 1996).

On the other hand, the crop markets in the shape of relative prices and transaction costs have induced agricultural intensification in Mattanur-Iritty, and encouraged crop shifts away from environment-friendly cashew cultivation. However, these processes have only a slight impact on the environment because they are moderate and because cultivators apply appropriate rubber- and coconut-cultivation technologies including soil-conservation measures. Generally, the extent and kind of agricultural intensification in Mattanur-Iritty does not affect the opportunities for future generations, and can thus be regarded as sustainable.

This case study may also challenge some common hypotheses regarding the effect of crop markets on sustainable development. For example, by showing that the influence of relative prices on the cropping pattern can be very limited, it contradicts neoclassical theory. On the other hand, some of my findings are consistent with the neoclassical-economist viewpoint. For instance, the concentration of cashew cultivation in North Kerala, encouraged by trade and "comparative advantage," is also appropriate from an environmental viewpoint. Furthermore, the cost of agro-chemicals discouraged cashew growers from applying environmentally unsound, intensive cultivation methods. However, those middle-class and rich cashew growers who choose not to engage in a more intensive, "chemicalized," land use are motivated more by their aversion to annoyance than by their concern about increased monetary outlays.

Although cashew is an export-oriented cash crop, its cultivation does not lead to environmental degradation, as community-based development and the Marxist-influenced viewpoint have suggested. Moreover, poor peasants were also able to benefit from the export-oriented raw-cashewnut production. Besides, this case study shows that poverty does not necessarily cause unsustainable cultivation practices. On the contrary, poverty may hinder peasants from shifting to ecologically less suitable crops than cashew, and to adopting less sustainable, or more intensive, cultivation methods. However, the example of the land redistribution to very poor, indebted people shows how the "simple reproduction squeeze" can come into force and result in unsustainable practices such as the clearing of cashew plantations.

Furthermore, the findings of this case study are consistent with indigenous-knowledge studies suggesting that farmers do not adopt whole packages of transferred technology developed in governmental agricultural-research stations because these packages may not be adapted to their situation. The high-yielding cashew varieties developed in research stations tend to be inappropriate on very poor soils, where cashew often grows in Mattanur-Iritty. Moreover, neither the cost-averse poor peasants nor the annoyance-averse cashew growers like the high-yielding variety of cashew because it requires more inputs and reduces the length of the cashew-harvest season, while failing to mitigate variation in yields and farming risks.

8 Markets For or Against Sustainable Development?

In view of the growing severity of environmental problems in many developing countries and the increased significance of markets in a globalizing economy, this study has attempted to explore the relationship between agricultural practice and sustainable development in a more complex and empirically grounded fashion than has been done to date. For too long, the debate has been polarized between simplistic support of "free" trade, on the one hand, and an anti-market stance on the other. Instead, this book argues for the importance of detailed, crop- and locality-specific research. The two selected case studies, alone, show how much the influence of crop markets on agricultural practice and on sustainable development may differ from one crop to another, and from one area to another, even in the same region. Detailed research on other places may reveal yet another picture. Moreover, it is important to stress that Kerala, where both research sites were located, is exceptional among Indian states, and certainly cannot be taken to represent developing countries in general. Nevertheless, this study does provide a number of general conclusions that may also have wider implications for theories of sustainable development and for environmental policy.

8.1 Crop Markets, Agricultural Practice and Development

The influence of crop markets on agricultural practice can vary widely, and individual access to resources, agrarian relations, state policies and technical-material factors also play an important role. The massive growth of pineapple cultivation in Vazhakulam, for example, was strongly connected with developments in the pineapple market. On the national level, the demand for pineapple has generally increased, thanks to India's growing urban middle class. Pineapple prices have risen comparatively quickly and transportation facilities have improved. However, not only these general market conditions but, particularly, the determination of traders in Vazhakulam to establish personal connections to fruit wholesalers in terminal markets also helped to lower transaction costs and to improve market access. Furthermore, cultivators in Vazhakulam improved pineapple-cultivation technologies in order to make full use of the new local marketing opportunities.

The cashew market, by contrast, had only a subsidiary impact on agricultural practice in Mattanur-Iritty. General market conditions and local marketing opportunities changed only gradually. Also, cashew growers have barely changed their practices: they did not react immediately to relative prices of cashew and competing crops or to changing state regulations regarding cashew marketing. Rather, they have continued to cultivate cashew because cashew cultivation is an unquestioned social practice, because it is most suitable under the particular bio-physical conditions of the locality, and because it involves low production costs and reduces the need to deal with other agents such as laborers, moneylenders or banks.

Evidence from studies of other crops in Kerala confirms that markets have only relative impact on agricultural practice. Rising prices and improved marketing support for rubber, for example, have encouraged middle-class and rich farmers to start new rubber plantations, but the comprehensive technical support of India's Rubber Board may have contributed just as much as agricultural markets to the expansion of rubber cultivation. Regarding paddy cultivation, furthermore, scarcity of paddy workers and regulated labor relations, rather than stagnating prices due to increased supply of rice from North India and state-administered prices, seem to have been the principal incentive for farmers to convert wetlands to other crops.

This study demonstrates that even in a commercialized, capitalist agricultural economy such as in Kerala, the role of social practices, agrarian relations and state interventions may be just as important as markets. Even when particular crop markets influence cropping patterns considerably, these markets still cannot be reduced to sheer price relations or demand-and-supply structures. In the case of new cash crops, for example, localized market access, market infrastructure and particular marketing arrangements between buyer and seller seem to be more significant determinants for agricultural land use than general market conditions. Particularly for poor cultivators, moreover, low cultivation costs appear to be more crucial than opportunities for high profits.

This implies that the effect of crop markets depends to a great extent on the pattern of interaction between buyer and seller, and on the way how markets are embedded in social structures and the state. The pineapple traders in Vazhakulam, for example, did not apply exploitative trading practices and offered prices that appeared "fair" to the cultivators. Outright malpractice by traders may have been prevented by the pineapple growers' economic and political power (rather than by any persistence of a "moral economy" among fellow villagers). Also, the traders did not interlock pineapple marketing with credit, land or labor markets, as they operate in a context of relatively well-developed formal credit facilities, formalized labor relations and legal prohibition of land tenancy.

On the other hand, the private traders in Vazhakulam were able to reduce exchange costs and to improve the access of local pineapple growers to terminal markets. Kerala's government, by contrast, failed to reduce transaction costs in its monopoly procurement of raw cashewnut. The fixed-price system even amplified farming risks because prices did not increase when supply decreased in years of bad harvests. Moreover, cashew growers in Mattanur-Iritty felt that the government favors the cashew-processing industry and factory workers, but discriminates against farmers by fixing low raw-cashewnut prices and by impeding profitable trade across the state border.

However, one cannot conclude that private marketing is invariably more efficient or more beneficial to farmers than state-controlled marketing. This study indicates that the state may reduce transaction costs, for example by defining and enforcing property rights or by providing market infra-structure; but it may also raise transaction costs through inappropriate regulation. More generally, the state is able to both advance equity at the local level and favoring dominant classes, making the effect of markets on the poor more benign or more damaging.

Arguably, only middle-class and rich cultivators in Vazhakulam were able to produce pineapple on a large scale and to benefit from the improved local pineapple-marketing opportunities. Because commercial pineapple cultivation requires investments that are beyond the means of poor peasants, the pineapple market did not become a medium of forced commerce and exploitation. The market-induced growth of pineapple cultivation had even a positive effect on poor households, as many benefited from additional employment opportunities and increased agricultural wages. However, these gains did not automatically "trickle down" to the poor, but were negotiated among various interest groups, including labor unions. The improved market access in Vazhakulam thus had a positive impact on local socioeconomic development.

The market-oriented cashew cultivation in Mattanur-Iritty, on the other hand, did not have much impact on local socioeconomic development, as cashew cultivation was only a subsidiary source of income for the farmers. Yet it did offer even poor farmers a chance to diversify cash income, because production costs are very low. Also for the poor, sale of cashewnut does not represent an instance of forced commerce, as they are offered the same price as big cashew growers, and they do not react to falling prices with increased labor input in order to increase production. In any case, labor intensity affects the level of cashew production only by a small amount; production depends more heavily on weather conditions.

In sum, these findings suggest that, in a situation of relatively equitable social and political structures, better market access tends to support socio-economic development, and gains from trade are likely to be distributed over various sections of society, even though not all groups may benefit equally.

8.2   Farmers, Markets and the Environment

Farmers in Kerala apply both environmentally friendly practices and environmentally unsound practices. Some significant processes, such as the spread of rubber monoculture, the overall "chemicization" of agriculture and, particularly, the widespread conversion of wetland paddy, may have fairly severe consequences for the environment (e.g., loss of biodiversity, pollution of soils and water, scarcity of water and more floods) that may threaten the livelihoods of vulnerable groups and future development opportunities. This indicates that Kerala's agriculture, though still compara-tively sustainable, is moving toward less sustainable development.

Despite the high levels of education and social awareness in Kerala, farmers are not primarily "ecologically concerned" producers, but rather in-dividuals with socioeconomic goals that they pursue both inside and outside agriculture. Neither environmentally friendly nor economically efficient farming is the farmers' first priority. Similarly, environmental degradation and agricultural stagnation do not seem to be so much the problems of farmers as of natural scientists, economists and a few social activists. In other words, farmers have not developed an "ecological sensitivity" to match that of environmentalists or even of some development planners. They do not seem to pass ethical judgement on inefficient or unsustainable resource use. In other words, the normative concept of sustainable development has not yet become a general cultural value in Kerala. This may be contrasted with the normative concept of social development or equity, which facilitated the successful implementation and enforcement of the land reform. The farmers' lack of commitment to the goals of sustain-able development that may also impede the sustainability-oriented community-based development initiatives that have recently been introduced in Kerala.

In regard to crop markets, this study indicates that they can affect the environment in very different ways. The localized pineapple-marketing opportunities, for example, had a concentrating effect in Vazhakulam, encouraging the spread of environmentally unsound pineapple cultivation into former paddy land in place of environmentally sound, but non-expandable pineapple intercropping on rubber replantations. Although this market-induced process undoubtedly had negative effects on the environment, it is wrong to blame the market for generating these unsustainable agricultural practices, as pineapple traders and consumers did not promote specific unsustainable cultivation methods. These practices were rather the consequence of the farmer's own low environmental awareness and lack of adequate knowledge about commercial pineapple cultivation. Rather, the market – i.e., the extensive use of human-made comparative advantage of pineapple cultivation in that region – perpetuated the application of inappropriate technology.

By contrast, market-oriented agriculture on the basis of ecological comparative advantage led to a environmentally sustainable concentration of cashew cultivation in Mattanur-Iritty and elsewhere in North Kerala, where climate and soils are more suitable for the cultivation of this cash crop than for the extensive cultivation of food crops. Cashew trees are effective in checking soil erosion, and farmers generally applied environmentally sound cultivation methods. In recent years, moreover, market-induced processes have been moderate in that region and have had only a minor impact on the local environment, with the exception of tightened import regulations in industrialized countries that had a major impact in discouraging the use of very harmful pesticides in cashew production.

These findings imply that market-induced growth often enhances negative environmental effects created by other factors such as lack of adequate knowledge about ecologically sound technologies, deficient state interventions, unequal access to resources and ill-defined property rights. In the case of rubber, for example, it was not the flourishing crop market but rather India's Rubber Board that was responsible for both encouraging the environmentally unsound practice of growing rubber in monoculture and promoting the environment-friendly building of terraces on rubber plantations. Furthermore, trade generally necessitates more transportation, which entails higher energy consumption and the generation of more waste.

This study also shows that commercialization can lead to specialized and intensified agricultural production, thereby reducing biodiversity as well as income diversity of farmers, and encouraging environmentally unsustainable agricultural practices as well as higher capital intensity of agricultural production. On the other hand, market-oriented agriculture can sometimes make possible regionally concentrated cultivation of cash crops in areas where the cultivation of food crops would be environmentally less sustainable.

As understood in terms of interactions between buyer and seller, the crop markets presented here did not promote unsustainable agricultural practices or unsustainable development, in general. Indeed, this book shows that markets even have the potential to direct agricultural producers toward more sustainable ways. However, it is not "free" trade but appropriate regulation of markets, pressure by consumer groups and "ecologically/socially concerned" consumption (C.N. Smith 1990) that can lead to more sustainable development in the Third World.

8.3   Implications for Theory and Future Research

The case studies provided empirical material that both confirms and contradicts the neoclassical-economist, Marxist-influenced and institutional viewpoints presented in Section 3.6. The institutional viewpoint seems to be the most adequate, indicating that crop markets can work both for and against sustainable development. The effect of markets depends much on the wider socioeconomic and technical-material context in which they are embedded, and on the pattern of interaction between buyer and seller. Therefore, neither presumption about markets and sustainable development held by neoliberal advocates of the "free" market, on the one hand, and by some environmentalists on the other, is adequate in scope.

Furthermore, factors often cited as the principal causes of environmental degradation in developing countries (e.g., "wrong" prices, unclearly defined property rights, unequal access to resources, asymmetric power relations, poverty, inappropriate "modern" technology) may be relevant in some cases but irrelevant in others. Instead of searching for yet another "leading cause" of unsustainable development, I suggest looking at the complex combinations of factors that contribute to underdevelopment and environmental degradation. This book maintains that it would be more useful to expand theoretically informed empirical research of actual problems of underdevelopment and environmental degradation in specific contexts. Such a research program would involve due consideration of local diversity and of locality-specific solutions. Therefore, it is also important to recognize the development potential of indigenous knowledge, institutions and community-based organizations within the wider context of economic and sociopolitical constraints and opportunities.

The nondeterministic concepts of social practice and real markets as well as qualitative methods used in this study hold much promise as a methodology for analyzing such context-specific development issues. Generally, research with qualitative methods tells us about farmers' perspectives, motives and motivations – information that is particularly important for participatory development planning and policy-making. For problem-oriented research, furthermore, specific examples of unsustainable agricultural practice could form a useful starting point. In order to measure the environmental consequences of these agricultural practices, natural science methods should be used in addition to qualitative and quantitative social science methods.

To complement the investigation of the relationship between crop markets and sustainable development, this book recommends a number of other research topics. For example, one may ask why social structures are more benign in some societies than in others for market-oriented development, or why agricultural markets for some products hinder the emergence of forced commerce more than others, because the consequences of commercialization seem to vary depending on the social context in which markets are embedded. To answer these questions would require careful comparative research and an examination of exactly how social structures change and are changed, and how exploitation through agricultural markets is constructed and broken. The findings of such a study could hold important lessons for policy-making.

The impact of India's economic liberalization and political decentralization on sustainable development forms another increasingly relevant research topic. As these policy trends coexist at present, it would be useful to investigate whether globalization and structural adjustment will simply "displace previous popular development initiatives at the grass-roots level" (Törnquist 1995: 91) or whether a retreating state and a liberalizing economy may also create scope for community-based sustainable development and for the better utilization of ecological and economic comparative advantage based on peculiar local endowment and indigenous knowledge. The latter hypothesis would suggest that globalization is not just a homogenizing process but, rather, one of "glocalization" – a concept that considers the relevance of local factors in the general process of globalization (see Robertson 1995).

8.4   Market-Based Policy Instruments

To conclude, I will recap ideas on the general usefulness of market-based policy instruments to achieve more sustainable development. We have seen that particular non-price measures related to the crop market may be an effective means to direct agricultural production toward more sustainability. These measures may include:

- trade restrictions on commodities that are produced with environmentally unsound methods or that contain residues from agrochemicals;
- pressure on states by consumer groups to implement of such trade restrictions;
- voluntary, "ecologically concerned" consumption.

These approaches involve state regulation of markets as well as community-based action of consumers. Furthermore, price incentives could be effective since farmers generally respond to changing relative prices, at least, in the long run. However, the internalization of environmental costs does not always seem practicable: Crop markets are often unable to distinguish between products that have been produced in sustainable ways and those that have not (e.g., between pineapples that have been grown on a rubber replantation and those that have been produced on a former paddy field). Moreover, taxation of agricultural inputs can have ambiguous effects, because the environmental impact may be:

- disproportionate to the quantity of the input used;
- dependent on characteristics of the local ecosystem;
- contingent upon the time in which the input is applied.

Price incentives may also be unsuitable for regional development planning, because markets tend to perpetuate concentration in areas with comparative advantage. Finally, environmental taxation or subsidy cuts may have negative socioeconomic consequences, particularly on poorer sections of society that rely on subsidized food and agricultural inputs.

In some but not all cases, there is a trade-off between socioeconomic development and environmental sustainability that cannot be overcome by technical solutions. Consequently, priorities need to be set. In doing so, I believe that the voices and the well-being of the poor must be given first priority.

Bibliography

Adams, B. (1993), "Sustainable Development and the Greening of Development Theory," in F.J. Schuurman (ed), *Beyond the Impasse*, Zed Books, London, 207-222.

Adams, J. (1986), "Peasant Rationality," *World Development*, 14(2), 273-282.

Adams, W.M. (1990), *Green Development*, Routledge, London.

Agarwal, A. and Narain, S. (1993), "Towards Green Villages," in W. Sachs (ed), *Global Ecology*, Zed Books, London, 242-256.

Agrawal, A.N. (1996), *Indian Economy*, 22nd Revised Edition, Wishwa Prakashan, New Delhi.

Alexander, W.M. (1994), "Exceptional Kerala," *Gaia*, 3(4), 211-226.

Anderson, T.L. and Leal, D.R. (1991), *Free Market Environmentalism*, Pacific Research Institute for Public Policy, San Francisco.

Arts, B. (1994), "Nachhaltige Entwicklung," *Peripherie*, 54, 6-27.

Athreya, V.B., Djurfeldt, G. and Lindberg, S. (1990), *Barriers Broken*, Sage, New Delhi.

Bailey, R. (1993), *Eco-Scam*, St. Martin's, New York.

Balassa, B. (1982), "Structural Adjustment Policies in Developing Economies," *World Development*, 10(1), 23-38.

Barbier, E. (1989), "Cash Crops, Food Crops, and Sustainability," *World Development*, 17(6), 879-895.

Bardhan, P. (1989), "The New Institutional Economics and Development Theory," *World Development*, 17(9), 1389-1395.

Barnes, T. and Gregory, D. (eds) (1997), *Reading Human Geography*, Arnold, London.

Bergh, J.C.J.M. Van Den (1996), *Ecological Economics and Sustainable Development*, Edward Elgar, Cheltenham.

Bernstein, H. (1982), "Notes on Capital and Peasantry," in J. Harriss (ed), *Rural Development*, Hutchinson University Library, London, 160-177.

Bernstein, H. (1990), "Taking the Part of Peasants?" in H. Bernstein *et al.* (eds), *The Food Question*, Earthscan, London, 69-79.

Berry, S.S. (1980), "Decision Making and Policymaking in Rural Development," in P.F. Barlett (ed), *Agricultural Decision Making*, Academic Press, New York, 321-335.

Bhaduri, A. (1973), "A Study in Agricultural Backwardness Under Semi-Feudalism," *Economic Journal*, 83, 120-137.

Bhaduri, A. (1986), "Forced Commerce and Agrarian Growth," *World Development*, 14(2), 267-272.

Bhalla, G.S. (ed) (1994), *Economic Liberalization and Indian Agriculture*, Institute for Studies in Industrial Development, New Delhi.

Bharadwaj, K. (1982), "Production Conditions in Indian Agriculture," in J. Harriss (ed), *Rural Development*, Hutchinson University Library, London, 269-288.

Bharadwaj, K. (1985), "A Note on Commercialization in Agriculture," in K.N. Raj *et al.* (eds), *Essays on the Commercialization of Indian Agriculture*, Centre for Development Studies and Oxford University Press, Trivandrum.

Bifani, P. (1992), "Environmental Degradation in Rural Areas," in A.S. Bhalla (ed), *Environment, Employment and Development*, ILO, Geneva, 99-120.

Blaikie, P. (1985), *The Political Economy of Soil Erosion in Developing Countries*, Longman, London.

Blaikie, P. (1995), "Understanding Environmental Issues," in S. Morse and M. Stocking (eds), *People and Environment*, UCL Press, London, 4-30.

Blaikie, P. and Brookfield, H. (1987), *Land Degradation and Society*, Methuen, London.

Blomström, M. and Hettne, B. (1984), *Development Theory in Transition*, Zed Books, London.

Boef, W. de, Amanor, K., Wellard, K. with Bebbington, A. (eds) (1993), *Cultivating Knowledge*, Intermediate Technology Publications, London.

Bohle, H.-G. (1990), "Von der Agrargeographie zur Nahrungsgeographie?" in B. Mohr *et al.* (eds), *Räumliche Strukturen im Wandel*, Teil B, Institut für Physische Geographie, Freiburg i. Br., 11-25.

Bolch, B. and Lyons, H. (1993), *Apocalypse Not*, Cato Institute, Washington D.C.

Booth, D. (1985), "Marxism and Development Sociology," *World Development*, 13(7), 761-787.

Booth, D. (ed) (1994), *Rethinking Social Development*, Longman, Essex.

Brokensha, D.W., Warren, D.M. and Werner, O. (eds) (1980), *Indigenous Knowledge Systems and Development*, University Press of America, Lanham.

Bryant, R.L. (1992), "Political Ecology," *Political Geography*, 11(1), 12-36.

Bryant, R.L. and Bailey, S. (1997), *Third World Political Ecology*, Routledge, London.

Cashew Bulletin (various issues), Cashew Export Promotion Council of India, Cochin.

Cashew Export Promotion Council (1994), *Indian Cashews*, Government of India, Cochin.

Centre for Development Studies (1977), *Poverty, Unemployment and Development Policy*, Orient Longman, Bombay.

Centre for Earth Science Studies (1984), *Resource Atlas of Kerala*, Government of India, Trivandrum.

Centre for Science and Environment (1985), *The State of India's Environment 1984-85*, New Delhi.

Chambers, R. and Conway, G.R. (1992), "Sustainable Rural Livelihoods," Discussion Paper 296, Institute of Development Studies, Sussex.

Chambers, R., Pacey, A. and Thrupp, L.A. (eds) (1989), *Farmer First*, Intermediate

Technology Publications, London.

Chattopadhyay, S. (1984), *Deforestation in Parts of Western Ghats Region (Kerala), India*, Centre of Earth Science Studies, Trivandrum.

Chayanov, A.V. (1966 [1925]), "Peasant Farm Organization," in D. Thorner *et al.* (eds), *A.V. Chayanov on the Theory of Peasant Economy*, The American Economic Association, Homewood, 29-317.

Collins, J. (1992), "Marxism Confronts the Environment," in S. Ortiz and S. Lees (eds), *Understanding Economic Process*, University Press of America and Society for Economic Anthropology, Lanham, 179-188.

Committee on Agro-Climatic Zones and Cropping Patterns (1974), *Report*, Government of Kerala, Trivandrum.

Costanza, R. (ed) (1991), *Ecological Economics*, Columbia University Press, New York.

Datt, R. and Sundharam, K.P.M. (1995), *Indian Economy*, 32nd Revised Edition, S. Chand, New Delhi.

Deepa, G.L. (1994), "Industrial Crisis and Women Workers," Unpublished M.Phil. Thesis, Centre for Development Studies, Trivandrum.

Devlin, J.F. and Yap, N.T. (1994), "Structural Adjustment Programmes and the UNCED Agenda," in C. Thomas (ed), *Rio: Unravelling the Consequences*, Frank Cass, Ilford, 63-79.

Directorate of Cashewnut Development (1992), *Centrally Sponsored Cashew Development Programmes in 8th Plan (1992-1997)*, Government of India, Cochin.

Directorate of Economics and Statistics (1992), *Report on Cost of Cultivation of Important Crops in Kerala 1989-90*, Government of Kerala, Trivandrum.

Directorate of Economics and Statistics, *Agricultural Statistics in Kerala*, various issues, Government of Kerala, Trivandrum.

Directorate of Economics and Statistics, *Season and Crops Report*, various issues, Government of Kerala, Trivandrum.

Directorate of Economics and Statistics, *Statistics for Planning*, various issues, Government of Kerala, Trivandrum.

Doeleman, J.A. (1992), "On Instruments of Environmental Policy," in A.S. Bhalla (ed), *Environment, Employment and Development*, ILO, Geneva, 79-98.

Drèze, J. and Sen, A. (1995), *India: Economic Development and Social Opportunity*, Oxford University Press, Delhi.

Eapen, K.V. (1986), *A Study of Kerala History*, Kollet Publication, Kottayam.

Eapen, M. (1994), "The Changing Structure of the Workforce in Kerala," in B.A. Prakash (ed), *Kerala's Economy*, Sage, New Delhi, 61-77.

Ehrlich, P.R. and Ehrlich, A.H. (1990), *The Population Explosion*, Simon & Schuster, New York.

Ellis, F. (1988), *Peasant Economics*, Cambridge University Press, Cambridge.

Ensminger, J. (1992), *Making a Market*, Cambridge University Press, Cambridge.

Evers, H.-D. (1987), "Subsistenzproduktion, Markt und Staat," *Geographische Rundschau*, 39(3), 137-140.

Evers, H.-D. and Schrader, H. (eds) (1994), *The Moral Economy of Trade*,

Routledge, London.

Extoxnet (Extension Toxicology Network) (1993), *A Universitary Pesticide Information Project*, http://ace.orst.edu/info/extoxnet, 23.6.1997.

FAO (Food and Agricultural Organisation of the United Nations) (1990-96), *FAOSTAT Agriculture Statistics Database*, http://www.fao.org/waicent, 29.6.1997.

Farm Information Bureau, *Farm Guide*, various issues, Government of Kerala, Trivandrum.

Farrington, J., Lewis, D.J., Satish, S. and Miclat-Teves, A. (eds) (1993), *Non-Governmental Organizations and the State in Asia*, Routledge, London.

FitzSimmons, M. (1997), "The Matter of Nature," in T. Barnes and D. Gregory (eds), *Reading Human Geography*, Arnold, London, 211-229.

Flick, U. (1995), *Qualitative Forschung*, Rowohlt, Hamburg.

Frank, A.G. (1969), *Capitalism and Underdevelopment in Latin America*, Monthly Review Press, New York.

Franke, R.W. and Chasin, B.H. (1994), *Kerala: Development Through Radical Reform*, Promilla, New Delhi.

Gadgil, M. and Guha, R. (1992), *This Fissured Land*, Oxford University Press, Delhi.

George, K.K. (1993), *Limits to Kerala Model of Development*, Centre for Development Studies, Trivandrum.

George, M. (1991), *Fertilizer Consumption and Agricultural Development in a Developing Economy*, Classical Publishing, New Delhi.

George, P.S. (1995), "State Interventions Relevant to Land Use," Manuscript, Centre for Development Studies, Trivandrum.

George, P.S. (1996), "Land Use and Cropping Pattern in Kerala," Manuscript, Centre for Development Studies, Trivandrum.

Ghai, D. (1994), "Environment, Livelihood and Empowerment," in D. Ghai (ed), *Development and Environment*, Blackwell and UNRISD, Oxford.

Ghai, D. and Vivian, J.M. (eds) (1992), *Grassroots Environmental Action*, Routledge, London.

Gibbon, D., Lake, A. and Stocking, M. (1995), "Sustainable Development," in S. Morse and M. Stocking (eds), *People and Environment*, UCL Press, London, 31-68.

Giddens, A. (1979), *Central Problems in Social Theory*, Macmillan, London.

Giddens, A. (1984), *The Constitution of Society*, University of California Press, Berkeley.

Glaeser, B. (ed) (1987), *The Green Revolution Revisited*, Allen & Unwin, London.

Gorz, A. (1980), *Ecology as Politics*, Black Rose Books, Montréal.

Govindaru, V. (1996), "Crop Selection, Farm Management and Livelihood," Manuscript, Centre for Development Studies, Trivandrum.

Grabowski, R. (1988), "The Theory of Induced Institutional Innovation," *World Development*, 16(3), 385-394.

Gulati, A. (1996), "Harvesting the Crop," *Economic and Political Weekly*, 31(15), 929-930.

Harriss, B. (1989), "Agricultural Merchants' Capital and Class Formation in India," *Sociologia Ruralis*, 29(2), 166-179.

Harriss, B. (1991), "Markets, Society and the State," Report to the World Institute for Development Economics Research, Helsinki.

Harriss, B. (1993), "Real Foodgrains Markets and State Intervention in India," in C. Hewitt de Alcántara (ed), *Real Markets*, Frank Cass, EADI and UNRISD, London, 61-81.

Harriss-White, B. (1996a), *A Political Economy of Agricultural Markets in South India*, Sage, New Dehli.

Harriss-White, B. (1996b), "Free Market Romanticism in an Era of Deregulation," *Oxford Development Studies*, 24(1), 27-45.

Harriss, J. (ed) (1982), *Rural Development*, Hutchinson University Library, London.

Harriss, J. (1994), "Between Economism and Post-Modernism," in D. Booth (ed), *Rethinking Social Development*, Longman, Essex, 172-196.

Hayami, Y. and Ruttan, V.W. (1985), *Agricultural Development*, Johns Hopkins University Press, Baltimore.

He, Y. (1994), "Economie Néo-Institutionnelle et Développement," *Revue d'Economie du Développement*, 4, 3-34.

Herring, R. (1983), *Political Economy of Agrarian Reform in South Asia*, Oxford University Press, Delhi.

Herring, R. (1989), "Dilemmas of Agrarian Communism," *Third World Quarterly*, 11(1), 89-115.

Hewitt de Alcántara, C. (ed) (1993), *Real Markets*, Frank Cass, EADI and UNRISD, London.

Indian Cashew Journal (various issues), Cashew Export Promotion Council of India, Cochin.

Indian Express (19.4.1995), "Cashew Procurement Act Quashed."

Isaac, T.M.T. (1994), "The Trend and Pattern of External Trade of Kerala," in B.A. Prakash (ed), *Kerala's Economy*, Sage, New Delhi, 368-393.

Iyer, S.R.R. (1996), "Social Development in Kerala, India," Unpublished M.Phil. Thesis, University of Hong Kong.

Jacob, J. (1996), "Institutional Interventions in Water Management," Manuscript, Centre for Development Studies, Trivandrum.

Jacobs, M. (1994), "The Limits of Neoclassicism," in M. Redclift and T. Benton (eds), *Social Theory and the Global Environment*, Routledge, London, 67-91.

Janvry, A. de, Sadoulet, E. and Thorbecke, E. (eds) (1993), "State, Market, and Civil Organizations," *World Development*, 21(4), 565-689.

Jose, D. (1991), "Homegardens of Kerala," Unpublished M.Sc. Thesis, Agricultural University of Norway.

Kannan, K.P. (1981), *Cashew Development in India*, Centre for Development Studies, Trivandrum.

Kannan, K.P. and Pushpangadan, K. (1988), "Agricultural Stagnation in Kerala," *Economic and Political Weekly*, 23(39), A120-A128.

Karshenas, M. (1994), "Environment, Technology and Employment," *Development and Change*, 25(4), 723-756.

Kerala Agricultural University (1993), *Package of Practices Recommendations*, Government of Kerala, Mannuthy.

Kerala State Gazetteer (1986), Vol. 1, Government of Kerala, Trivandrum.

Kerm, P. Van (1997), "From 'State Domestic Product' to 'Real Income'," Manuscript, University of Namur.

Kirchner, J., Ledec, G., Goodland, R. and Dake, J. (1984), "Carrying Capacity, Population Growth, and Sustainable Development," Staff Working Paper 690, World Bank, Washington D.C.

Kohli, A. (1989), "Politics of Economic Liberalization in India," *World Development*, 17(3), 305-328.

Kothari, M. and Kothari, A. (1993), "Structural Adjustment vs Environment," *Economic and Political Weekly*, 28(11), 473-477.

Kothari, S. and Parajuli, P. (1993), "No Nature Without Social Justice," in W. Sachs (ed), *Global Ecology*, Zed Books, London, 224-241.

Krishnaswamy, K.S. (1994), "Agricultural Development Under the New Economic Regime," *Economic and Political Weekly*, 29(26), A65-A71.

Kurien, C.T. (1994a), "Kerala's Development Experience," *International Congress on Kerala Studies*, Abstracts, Vol. 1, AKG Centre for Research and Studies, Trivandrum.

Kurien, C.T. (1994b), *Global Capitalism and the Indian Economy*, Orient Longman, New Delhi.

Kurup, K.K.N. (1981), *William Logan: A Study in the Agrarian Relations of Malabar*, Sandhya Publications.

Kuttikrishnan, A.C. (1994), "Educational Development in Kerala," in B.A. Prakash (ed), *Kerala's Economy*, Sage, New Delhi, 349-367.

Laclau, E. (1971), "Feudalism and Capitalism in Latin America," *New Left Review*, 67.

Lal, D. (1983), *The Poverty of 'Development Economics'*, Institute of Economic Affairs, London.

Lal, D. (1990), "Political Economy and Public Policy," Occasional Paper 19, International Center for Economic Growth, San Francisco.

Land Use Board (1976), *Identification of Compact Vacant Areas Suitable for Cashew Cultivation in Cannanore District*, Government of Kerala, Trivandrum.

Lélé, S.M. (1991), "Sustainable Development," *World Development*, 19(6), 607-621.

Lipton, M. (1968), "The Theory of the Optimising Peasant," *Journal of Development Studies*, 4(3), 327-351.

Little, P.D. and Horowitz, M.M. (eds) (1987), *Lands at Risk in the Third World*, Westview Press, Boulder.

Long, N. and Ploeg, J.D. van der (1994), "Heterogeneity, Actor and Structure," in D. Booth (ed), *Rethinking Social Development*, Longman, Essex, 62-89.

Mackintosh, M. (1990), "Abstract Markets and Real Needs," in H. Bernstein *et al.*

(eds), *The Food Question*, Earthscan, London, 43-53.

Mann, S. (1990), *Agrarian Capitalism in Theory and Practice*, University of North Carolina Press, Chapel Hill.

Market Asia Index (1996), *Nuts About Organics*, http://www.milcom.com/rap/index.html, 23.6.1997.

Martell, L. (1994), *Ecology and Society*, Polity Press, Cambridge.

Mathew, E.T. (1995), "Unemployment and Self-Employment," *Economic and Political Weekly*, 30(44), 2815-2826.

Maurya, D.M. (1989), "The Innovative Approach of Indian Farmers," in R. Chambers *et al.* (eds), *Farmer First*, Intermediate Technology Publications, London, 9-14.

McKibben, B. (1995), *Hope, Human and Wild*, Little, Brown & Co, Boston.

Meadows, D., Meadows, D., Randers, J. and Behrens, W. (1972), *The Limits to Growth*, Universe, New York.

Menon, D. (1994), *Caste, Nationalism and Communism in South India*, Cambridge University Press, Cambridge.

Mies, M. and Shiva, V. (1993), *Ecofeminism*, Zed Books, London.

Mohandas, M. (1994), "Poverty in Kerala," in B.A. Prakash (ed), *Kerala's Economy*, Sage, New Delhi, 78-94.

Mohandas, M. (1995), *Farmer Friendly Marketing of Fruits and Vegetables*, Kerala Horticulture Development Programme, Trivandrum.

Müller-Böker, U. (1995), "Ethnoökologie," *Geographische Rundschau*, 47(6), 375-379.

Nair, G.S. (1983), "Agricultural Statistics in Kerala," Unpublished Internal Paper of the Director, Directorate of Economics and Statistics, Government of Kerala.

Nair, G.P.R. (1994), "Migration of Keralites to the Arab World," in B.A. Prakash (ed), *Kerala's Economy*, Sage, New Delhi, 95-114.

Nair, G.P.R. and Pillai, M.P. (1994), *Impact of External Transfers on the Regional Economy of Kerala*, Centre for Development Studies, Trivandrum.

Nair, K.N., Narayana, D. and Sivanandan, P. (1989), *Ecology or Economics in Cardamom Development*, Centre for Development Studies, Trivandrum.

Narayanan, N.C. (1996), "Land Use Interventions in Fragile Slopes of Kerala and Sustainable Livelihoods," Manuscript, Centre for Development Studies, Trivandrum.

North, D.C. (1990), *Institutions, Institutional Change and Economic Performance*, Cambridge University Press, Cambridge.

Nossiter, T.J. (1982), *Communism in Kerala*, Hurst, London.

Nossiter, T.J. (1988), *Marxist State Governments in India*, Pinter, London.

O'Connor, J. (1988), "Capitalism, Nature, Socialism," *Capitalism, Nature, Socialism*, 1, 11-38.

Ohler, J.G. (1979), *Cashew*, Koninklijk Instituut voor de Tropen, Amsterdam.

Oommen, M.A. (1993), *Essays on Kerala Economy*, Oxford & IBH, New Delhi.

Oommen, M.A. (1994), "Land Reforms and Economic Change," in B.A. Prakash (ed), *Kerala's Economy*, Sage, New Delhi, 117-140.

Parayil, G. (1996), "The 'Kerala Model' of Development," *Third World Quarterly,* 17(5), 941-957.

Paul, A. (1996), "Management of Backwater Resources and Sustainable Livelihoods in Kerala," Manuscript, Centre for Development Studies, Trivandrum.

Pearce, D. (1988), "The Sustainable Use of Natural Resources in Developing Countries," in R.K. Turner (ed), *Sustainable Environmental Management,* Belhaven Press, London, 102-117.

Pearce, D., Barbier, E., Markandya, A., Barrett, S., Turner, R.K. and Swanson, T. (1991), *Blueprint 2,* Earthscan, London.

Pearce, D., Markandya, A. and Barbier, E. (1990), *Sustainable Development,* Earthscan, London.

Peet, R. and Watts, M. (1993), "Introduction: Development Theory and Environment in an Age of Market Triumphalism," *Economic Geography,* 69(3), 227-253.

Pillai, P.P. (1994), *Kerala Economy,* Institute of Planning and Applied Economic Research and John Matthai Foundation, Thrissur.

Polanyi, K. (1968), "The Economy as Instituted Process," in E.E. LeClair and H.K. Schneider (eds), *Economic Anthropology,* Holt, Rinehart and Winston, New York, 122-143.

Popkin, S.L. (1988), "Public Choice and Peasant Organization," in R.H. Bates (ed), *Toward a Political Economy of Development,* University of California Press, Berkeley, 245-271.

Prakash, B.A. (1987), *Agricultural Development of Kerala from 1800 AD to 1980 AD,* Centre for Development Studies, Trivandrum.

Prakash, B.A. (ed) (1994), *Kerala's Economy,* Sage, New Delhi.

Pronk, M. (1997), "Clay Mining and Brick Production in Trichur District," Manuscript, University of Zurich.

Py, C., Lacoeuilhe, J.J. and Teisson, C. (1984), *L'Ananas,* Edition G.-P., Maisonneuve and Larose, Paris.

Rai, V. (1988), "Marketing and Processing of Agricultural Produce in Kerala," in *Five Year Plan Workshop,* Government of Kerala, Trivandrum, 252-261.

Raj, K.N., Bhattacharya, N., Guha, S. and Padhi, S. (eds) (1985), *Essays on the Commercialization of Indian Agriculture,* Centre for Development Studies and Oxford University Press, Trivandrum.

Randhawa, N.S. (1994), "Liberalization and Implications for Agricultural Policy," in G.S. Bhalla (ed), *Economic Liberalization and Indian Agriculture,* Institute for Studies in Industrial Development, New Delhi, 353-390.

Rao, H.C.H. (1994), *Agricultural Growth, Rural Poverty and Environmental Degradation in India,* Oxford University Press, Delhi.

Redclift, M. (1987), *Sustainable Development: Exploring the Contradiction,* Methuen, London.

Redclift, M. (1994), "Sustainable Development: Economics and the Environment," in M. Redclift and C. Sage (eds), *Strategies for Sustainable Development,* John Wiley, Chichester, 17-34.

Redclift, M. and Benton, T. (eds) (1994), *Social Theory and the Global Environment*, Routledge, London.

Reed, D. (ed) (1992), *Structural Adjustment and the Environment*, Westview Press, Boulder.

Reichart, T. (1982), *Die Ananas*, Wirtschafts- und Sozialgeographisches Institut der Friedrich-Alexander-Universität, Nürnberg.

Rhoades, R. (1989), "The Role of Farmers in the Creation of Agricultural Technology," in R. Chambers *et al.* (eds), *Farmer First*, Intermediate Technology Publications, London, 3-9.

Richards, P. (1985), *Indigenous Agricultural Revolution*, Hutchinson, London.

Robertson, R. (1995), "Glocalization," in M. Featherstone *et al.* (eds), *Global Modernity*, Sage, London, 25-44.

Sachs, W. (ed) (1993), *Global Ecology*, Zed Books, London.

Salam, A.M. and Mohanakumaran, N. (1996), "Towards a More Sustainable Cashew Industry in India," *Cashew Bulletin*, 23(6), 3-11.

Santhakumar, V. (1995), "Research on Sustainable Agriculture Compared," *ILEIA Newsletter*, 11(2), 24-25.

Saradamoni, K. (1983), "Changing Land Relations and Women," in R. Mehra and K. Saradamoni, *Women and Rural Transformation*, Concept Publishing, New Delhi, 33-171.

Schultz, T.W. (1964), *Transforming Traditional Agriculture*, Chicago University Press, Chicago.

Schutz, A. (1962), *Collected Papers*, Vol. 1 (edited by A. Brodersen), Martinus Nijhoff, The Hague.

Schuurman, F.J. (ed) (1993), *Beyond the Impasse*, Zed Books, London.

Scott, J.C. (1976), *The Moral Economy of the Peasant*, Yale University Press, New Haven.

Sen, A. (1981), *Poverty and Famines*, Clarendon Press, Oxford.

Sen, S.K. (1990), "Pineapple," in T.K. Bose (ed), *Fruits*, Naya Prokash, Calcutta, 252-279.

Sessions, G. (ed) (1995), *Deep Ecology for the 21st Century*, Shambhala, Boston.

Sivandandan, P.K. (1994), "Performance of Agriculture in Kerala," in B.A. Prakash (ed), *Kerala's Economy*, Sage, New Delhi, 141-159.

Smith, C.N. (1990), *Morality and the Market*, Routledge, London.

Smith, N. (1984), *Uneven Development*, Blackwell, Oxford.

So, A.Y. (1990), *Social Change and Development*, Sage, Newbury Park.

State Planning Board, *Economic Review*, various issues, Government of Kerala, Trivandrum.

Stiglitz, J.E. (1986), "The New Development Economics," *World Development*, 14(2), 257-265.

Sunanda, S. (1991), "Institutional Credit for Agriculture in Kerala," Unpublished M.Phil. Thesis, Centre for Development Studies, Trivandrum.

Suresh, K.A. and Joseph, M. (1990), *Public Participation in Rural Development*, Kerala Agricultural University, Mannuthy.

Swaminathan, M.S. (1993), "Farm Policy," *The Hindu: Survey of the Environment, 1993*, 28-29.

Swedberg, R. (1994), "Markets as Social Structures," in N.J. Smelser and R. Swedberg (eds), *The Handbook of Economic Sociology*, Princeton University Press, Princeton, 255-282.

Tharamangalam, J. (1984), "The Penetration of Capitalism and Agrarian Change in Southwest India, 1901-41," *Bulletin of Concerned Asian Scholars*, 16(1), 53-62.

The Hindu (19.4.1995), "HC Quashes Monopoly Procurement of Cashew."

Thomas, C. (ed) (1994), *Rio: Unravelling the Consequences*, Frank Cass, Ilford.

Törnquist, O. with Tharakan, P.K.M. (1995), *The Next Left?*, Nordic Institute of Asian Studies, Copenhagen.

Trade Development Authority (1986), *Action Plan for Promoting Export from Kerala*, Government of India, New Delhi.

Turner, R.K., Pearce, D. and Bateman, I. (1994), *Environmental Economics*, Harvester Wheatsheaf, New York.

Unnikrishnan, S. (1993), "To Many Dams Spell Death for a Kerala River," *Down To Earth*, 1(19), 46-47.

Varghese, T.C. (1970), *Agrarian Change and Economic Consequences*, Allied Publishers, Bombay.

Verghese, K.E. (1986), *Socio-Economic Change in Kerala*, Ashish, New Delhi.

Véron, R. (1993), "Kleinunternehmer-Selbsthilfegruppen im städtischen Java, Indonesien," Unpublished M.Sc. Thesis, University of Zurich.

Véron, R. (1997), "Pineapple-Related Land Management and Pineapple Markets in Kerala," Project Paper 1, Centre for Development Studies and Geographical Institute, University of Zurich, Trivandrum and Zurich.

Véron, R. (1998), "Markets, Environment and Development in South India," Unpublished Ph.D. Thesis, University of Zurich.

Visvanathan, S. (1986), "Reconstruction of the Past Among the Syrian Christians of Kerala," *Contributions to Indian Sociology*, 20(2), 241-260.

Volkamer, J.C. (ed) (1987 [1714]), *Continuation der Nürnbergischen Hesperidum*, Lithographie Hans Arndt, Nürnberg.

Warford, J. (1989), "Environmental Management and Economic Policy in Developing Countries," in G. Schramm and J. Warford (eds), *Environmental Management and Economic Development*, World Bank, Washington D.C., 7-22.

WCED (World Commission on Environment and Development) (1987), *Our Common Future*, Brundtland Report, Oxford University Press, New York.

Webster's Ninth New Collegiate Dictionary (1991), Merriam-Webster, Springfield.

Werlen, B. (1993), *Society, Action and Space*, Routledge, London.

World Resource Institute in collaboration with UNEP and UNDP (1994), *World Resources 1994-95*, Oxford University Press, New York.

Zachariah, M. and Sooryamoorthy, R. (1994), *Science for Social Revolution*, Zed Books, London.

Zimmermann, M.E. (1994), *Contesting Earth's Future*, University of California Press, Berkeley.

Index